THE SCIENCE OF OURSELVES

THE SCIENCE OF OURSELVES

A SCIENCE TO THE DESCENT OF MAN

BAMPFYLDE FULLER

GYAN

Published by

Gyan Publishing House
5, Ansari Road
Daryaganj, New Delhi-110002
Phone: 011-47034999, 9811692060
E-mail: books@gyanbooks.com

Distribution Network
gyanbooks.com
India, USA, Canada, UK, Australia

ISBN: 978-93-7169-074-4 (PB)

First Published, 1921

2nd Impression 2025

Printed at: Gyan Press, Delhi.

THE SCIENCE OF OURSELVES
Author: BAMPFYLDE FULLER

THE
SCIENCE OF OURSELVES

THE

SCIENCE OF OURSELVES

(A SEQUEL TO THE 'DESCENT OF MAN')

BY

SIR BAMPFYLDE FULLER

K.C.S.I., C.I.E.

AUTHOR OF 'STUDIES OF INDIAN LIFE AND SENTIMENT', 'THE EMPIRE OF INDIA', 'LIFE AND HUMAN NATURE', AND OTHER WORKS

1921

PREFACE

THERE are few, nowadays, who dispute the conclusion that the features of man's body have evolved from conditions approaching those of the ape, and that the physical characters of the ape have been slowly elaborated from a primitive form of life. If, however, this be true, the mind of man must also have evolved from a simpler condition. Its most salient feature is that of consciousness—that we should think before acting, and should know that we are thinking. If evolution is a reality, conscious life must have evolved from the unconscious—unless we endow a zoophyte with consciousness ; and this book endeavours to show how this may have come about. Its aim is to set out an inferential theory which will explain the origin and course of feeling, thought, and behaviour—an undertaking which must fall short of full achievement. But it will have been successful if it stimulates readers to think for themselves.

Our inquiries will take us off the ordinary course of scientific studies. For these, in general, occupy themselves with the classification of objects by their features, or with the quantitative appreciation of changes—in other words, they are more concerned with the definition of things and the measurement of consequences, than with the nature of the causes out of which they arise.

Velocity, for instance, is a consequence, which is quantitatively computed by establishing a ratio between units of space and units of time, as when stated to be 1,000 feet per second. Its cause remains obscure: nor is it necessary for the purpose of this calculation to justify the association in proportion of (apparently) such very different things as space and time, and the use of one to divide the other. Human nature produces consequences that can be measured arithmetically: we can deal statistically with the perceptible consequences of loyalty and disaffection, of industry and idleness. The object of this book is, however, to use reasoning inference as an instrument for detecting the causes from which such consequences spring. Behind these causes lies a chain of others which evade our reasoning powers, and lead us back to the problem of life itself. But it is something to have pushed back the screen—by however little.

We shall see that reasoning inference leads to the conclusion that our feelings and thoughts are nervous processes, owing their existence to evolutionary developments of the spinal cord and the brain. This is a view which it is very hard to accept. It does violence to our respect for the human mind, and for ourselves. It runs counter to our cherished opinions. But these are influenced by Sentiment, and of this Reason takes no account. It argues from generalized experiences of similarities and differences, of sequences and non-sequences, and obliges itself to accept the conclusions to which they lead. And Charity, at all events, should welcome an attempt to

PART I

NERVOUS ACTIVITIES

CHAPTER I

THE THREE PLANES OF NERVOUS ACTIVITY

If we were making a collection of the intricacies of nervous tissue which increase in complexity as we ascend the animal kingdom, we might arrange them in three classes, according as they were concerned with the automatic response to a stimulus, with the development of energy between the stimulus and the response, and with the development of conscious sensation and thought between the stimulus and the response. That is to say, nervous evolution has progressed through three different planes, constituting three different systems. Evolution is a process, not of destruction, but of development: it builds, so to speak, upon the roofs of old constructions, and the roofs remain, however deeply they may become embedded. Accordingly, we are obsessed by the proclivities of the remotest and humblest of our progenitors: our nervous equipment includes systems of all three descriptions; and, vaguely appreciating their differences, we term them the physical, the spiritual, and the mental—or, in other words, the instinctive, the psychic, and the reflective.

A nerve is set in action by a stimulus that impresses it. We think of stimuli as of things which affect us from the outside and are perceived through the senses. But our impressions of them afford little clue to their real character. Light and sound—two very potent stimuli—have been resolved into vibrations which are dark and noiseless, and our sensations of them are in reality their *consequences* upon the nerves. If, as is believed, the substantial objects that surround us are in a condition of vibration, our impressions of touch, such as those of rigidity and smoothness, are also the consequences of vibrations. Modern science is inclined to hold that the world of our environment is a congeries of heterogeneous vibrations. In this case the external stimuli that move us are peculiar classes of vibrations to which our sense organs are susceptible.

The primitive function of a nerve-system is to transform these vibrations into muscular movements—that is to say, it transforms movement of one kind into movement of another. Its action, accordingly, involves two different currents—one of sensory excitement from the sense-organ to the nerve-cell, the other of motor excitement from the nerve-cell to a muscle. This transformation of one kind of excitement into another kind must involve a change of neural condition in the nerve-cell. The nature [1] of *sensory* excitement is still obscure, but is believed to have affinities with an electro-motive current.[2] From the fact that *motor* excitement is primarily attended by mechanical consequences —that is to say, by muscular movement—we may perhaps assume that it is itself of a mechanical character, that it is a vibratory impulse, which takes one or other of two different phases, according as the muscular movement which it actuates is one of expansion, or approach, or of contraction, or recoil.

In the higher animals, nerves that activate muscles are *paired* systems, each comprising two cells [3]—one sensory, the other motor—clinging [4] to one another by

[1] 'Nothing has been ascertained regarding the nature of the change which is set up in the nerve element, and which travels along it in what is called the sensory impulse.' (Schafer's *Text-Book of Physiology*, ii. 606.)

[2] Nerves can be artificially thrown into a condition of electrical excitement, and muscular movement will follow as a response.

[3] 'The whole neural system may be regarded as a super-structure resting upon two nerve-cells—the two root-cells, one afferent, the other efferent. The afferent root-cell is the alpha of every functional series of units in the system: the efferent root-cell is the omega to which every series immediately or mediately leads.' (Professor Sherrington in Schafer's *Text-book of Physiology*, ii. 784.)

[4] Did the dendrons release hold, direct communication between the two cells would apparently be interrupted.

branchlets (dendrons) which spring from them. Longer branchlets (axons or fibrils) connect the sensory cell with its sense-organ, the motor-cell with a muscle. The intricate nerve-tissue of the spinal cord includes myriads of these pairs, and provides nerve-paths by which one cell can communicate with the other in roundabout fashion—through other parts of the cord or the brain—as well as directly. But this duality of composition—this distinctness of sensory and motor elements—is masked, and not clearly traceable, in the nervous tissue of the cerebellum and the brain.

Let us now revert to the three planes or systems of nervous organization. And, first, of that which we have distinguished as *physical.* This is also known as 'reflex' or 'instinctive'. A nervous impression is immediately and inevitably transformed into a movement which is appropriate to the impression. There is no room for any deliberate profiting by experience.[5] And (as we are assured by the highest authorities) brainless animals cannot be aware of the character of their impressions. But their sense-organs may, nevertheless, be very highly developed, as in the insects, lobsters, and cuttle-fish. They may possess exceedingly elaborate eyes, but they 'see' nothing in the ordinary sense of the word, and their eyes merely increase the complexity of the stimuli to which they automatically reply. That a nerve-system of this kind survives in man is proved by his 'reflex' actions. Coughing and winking, for instance, and the start of alarm, are automatic responses which are quite independent of consciousness. We may be aware of them, but our awareness follows and does not precede them. Indeed there may be violent responsive

[5] Henri Fabre illustrates the unintelligence of instinct by the conduct of a mason-wasp which he was watching in the construction of its nest. It had finished the little mud dome, leaving an aperture through which it had inserted its egg, with some spiders to provide the grub with food, and had flown away to procure mud for the cover. M. Fabre removed the nest and put it in his pocket. The insect returned, and contentedly plastered the cover upon the scar that the nest had left on the wall.

movements of repulsion during sleep, or when consciousness is deadened by an anaesthetic, as under the dentist's forceps or the surgeon's knife. That we can be awakened from sleep proves that we can be stimulated otherwise [6] than through the brain. For during sleep the brain is out of circuit. We can, moreover, feel that our appetites and emotions possess an insistency for which the brain does not account—that they are urged by stimuli which affect us instinctively as well as consciously. If their effect were not transmitted to the brain, their influence would be hidden from us, as is that of the stimuli upon which the internal functioning of the body depends. This, of course, remains purely instinctive, and in normal conditions is unattended by consciousness.

The second, or *psychic*, plane is marked by the development of energy, which goes to reinforce, transform, or prolong the response to a stimulus. We may infer that this accompanies the massing of nerve-cells into clusters or ganglia. Violent and sustained energy is manifested very strikingly by insects. Yet they show no traces of it in their larval stage. There is an extraordinary contrast between the sluggishness of a caterpillar and the activity of the butterfly into which it turns. It is a significant fact that this metamorphosis is attended by a concentration of nerve centres. Those of the caterpillar are distributed in a series through the segments of its body. In the butterfly [7] five of them coalesce to form one large complex ganglion. Let us make an assumption the rashness of which will be excused if we find, on proceeding, that it is borne out by other facts and is an adequate explanation of the accumulation of, so to speak, a store of energy. Let

[6] The nerve-fibrils that come from the sensory organs (the olfactory excepted) are not rooted in the cerebral hemispheres, and currents from them need not necessarily affect the brain.

[7] During the caterpillar's metamorphosis all its internal organs are re-cast. Newport, in tracing out these astonishing changes, noted the significant fact that the nerves were remodelled before the alimentary canal, generative system, and other organs. (Newport's *Art Insecta*, pp. 962–5.)

us suppose that certain particular currents of sensory excitement, on reaching these ganglia, are transformed into motor excitement which finds no muscular outlet and consequently expends itself in maintaining a group of nerve-cells in a condition of vibratory activity. This activity would take one or other of two phases according as the vibrations were stimulated by favourable or unfavourable impressions—according, that is to say, as, if translated into movement, their consequences would be movements of approach or of recoil. When these psychic vibrations impress the brain we are conscious of them as *feelings*. But they exist independently of consciousness in brainless animals.

In vertebrate animals the spinal cord, and the large spinal ganglion known as the *cerebellum*, are dense agglomerations of nerve-cells, and these animals manifest in more or less degree the oscillations of nervous energy that in man become the emotions and will. We know from personal experience that this energy not only reinforces the nervous effects of stimuli. It may neutralize or even reverse them, as, when moved by anger or pride, one refuses pleasures which would ordinarily attract him.

Nervous development culminates in the brain. This organ introduces two new elements into nervous life—ideas and consciousness.[8] It is peculiar to vertebrate animals but is not possessed by all of them. The humblest animal of this sub-kingdom—the Amphioxus—is brainless, and in the most primitive of fishes the brain is concerned with little but smell. That it is a distinct evolution from the spinal cord may be gathered from its embryonic development. The cerebral hemispheres

[8] Ferrier's *Functions of the Brain* (second edition, pp. 220 and 424). Professor Sherrington concludes that 'all purely spinal reactions fail to evince features of consciousness' (Schafer's *Text-book of Physiology*, ii. 860). In the same work (ii. 697) Sir Edward Schafer is explicit that 'the cerebral cortex is the seat of the intellectual functions, of intelligent sensation or consciousness, of idealism and of memory. That this is so is proved by the observation in man of cases of defective development, and of destructive diseases'.

are not represented in the primary nervous axis of the embryo, from which the spinal cord, the cerebellum, and the basal portion of the brain are evolved. They originate in a later vesicular outgrowth,[9] which develops, it seems, into a subsidiary nervous system that is interposed between the sensory susceptibilities of the body and their consequences in muscular action, and introduces a new and peculiar nervous condition—that of consciousness. Without the brain there is no definite consciousness. Animals may survive the excision of the brain. But they are reduced to the condition of complex machines.

When an impression of the sensory nerves reaches the brain it becomes a *sensation*. By inference we can analyse the sensation into two elements—one durable, the other transient. The first is its *character*, the second is its *conscious activity*. Its character persists when it passes from before us and is stored as a recollection. This is a record of a sensation, stereotyping its character, as a gramophone-record stereotypes the sounds that have been thrown upon it. And just as the gramophone record is silent until it is turned below the needle, so the recollection vanishes until it is recalled in thought. It then becomes 'illuminated' by the transient element of conscious activity. This is a feeling of agreeableness[10] or disagreeableness which is not fixed but variable. Impressions which please us greatly when we are in good spirits lose much of their glamour when we are irritated. Indeed, their agreeableness may be changed into disagreeableness. Music which pleases us in one mood may annoy us in another.

The existence of two elements in a sensation corresponds with the double nature of nervous excitement

[9] Bastian's *Brain as an organ of the Mind*, pp. 383, 384.

[10] 'All normal sensation, in due alternation and degree, is pleasurable' (Bain's *Emotions and Will*, p. 284). Ferrier concluded that sensations are accompanied by feelings which are divisible into two great and opposite classes—pains and pleasures (*Functions of the Brain*, p. 429). Cuvier remarks that every sensation is more or less agreeable or disagreeable (*Introduction to the Animal Kingdom*).

as sensory and motor. The character of the sensation represents that of the sensory excitement of which it is the brain-stamp, and this again results from the character of the stimulus. Its conscious agreeableness or disagreeableness would be accounted for did we extend to the brain the hypothesis that motor excitement, for which no muscular outlet exists, sets up vibratory nervous activity. Motor excitement must accompany sensory excitement, since one is the inevitable product, or metamorph, of the other. Consequently sensory excitement which reaches the brain must generate motor excitement, which will take one or other of two phases according as the sensory excitement is favourable or unfavourable. If this motor excitement discharged itself by communicating its activity to the brain-cell that stereotypes the sensory excitement from which it was derived, it would have the effect, we may suppose, of 'illuminating'[11] it—that is to say of rendering it 'conscious'. The consciousness is in fact ourselves. Accordingly we can say that an impression is conscious, or that we are conscious of an impression.

We are, then, affected by three kinds of motor excitement—one of the physical nervous system, another of the psychic, and the third of the brain. The first converts itself automatically into reflex movement and may be termed 'muscular'. The second is generated by sensory excitement which travels to the spinal cord and maintains the continuing energy of our waking hours. It is commonly distinguished as 'spiritual'. The third is 'cerebral', since it proceeds from sensory excitement which, running to the brain, stamps itself upon it. Cerebral motor excitement illumines the brain-stamp and renders it 'conscious'.

We ordinarily think of sensations as representing

[11] We can trace an analogy in the effect of light. Its vibrations, when they affect an object (or the vibrations of an object), give it a quality which, to the eyes, *illuminates* it.

things in our environment which are perceived by touch, smell, taste, sight, or hearing. But we also experience sensations of internal touch—that is to say, of self-touches. Pain is obviously internal touch—the pressure of deranged or injured tissue—since internal pain closely resembles that caused by external pressure. There is no difference of kind between the agony of gout and that caused by tight boots. Through internal touch,[12] moreover, we are sensible of the movements of our limbs, and can remember them: we have recollections of our actions and can think of moving our limbs without doing so. That we may be touched internally will not seem strange if it be realized that to the nervous system the muscles and organs of the body are external objects. There is a further development of sensation. We are consciously impressed by certain neural conditions and can remember them. For one system of nerves is external to the other. Hunger and alarm are neural conditions: they impress the brain and we form sensations of them as when we *feel* alarmed or hungry.

Moreover, we form sensations of psychic motor excitement. Pain is accompanied by a sensation of shrinking, or contracting; touches and movements of the limbs by sensations of effort and of resistance,[13] and we recollect these sensations. We remember the amount of effort that is needed to raise a particular object, and if we lift an empty jug which we believe to be full, we exert ourselves disproportionately and experience a feeling of surprise. We form sensations of our emotions and can remember a particular emotional experience, such as love or anger, in association with the recollection of the person who excited it. That we should form sensations of psychic energy is not an inexplicable mystery. For, if psychic excitement is of a vibratory nature, it might

[12] The joints and muscles are richly supplied with sensory nerves. The muscles are furnished with a network of them. Ferrier's *Functions of the Brain*, p. 68. See also Professor Sherrington in Schafer's *Text-book of Physiology*, ii. 1002.

[13] Hence we speak of actively touching an object as 'feeling' it.

act upon the internal sensory nerves like an impression of sight or sound, setting up sensory excitement which, on reaching the brain, would produce an impression of the psychic excitement,[14] and of its strength and kind—that is to say, of the amplitude and phase of its vibrations. There is no difficulty in believing that the sensory nerves can be stimulated by motor excitement directly and without the intervention of sense-organs. For, as we have seen, our sensations of internal touch must arise in this fashion.

Accordingly we receive sensations from two worlds—of our own muscular, neural, and psychic conditions, and of things that are outside us. We give the former the name of 'feelings'. They are, in fact, sensations of ourselves which the brain can present to us because it stands apart from the physical and psychic planes, and can be impressed by their conditions.

The brain is, moreover, a laboratory of complex ideas, and, as they are gradually perfected, becomes a symbolic miniature of our environment and ourselves. The materials upon which it works are recollections, which are transformed into ideas by being coupled together, or by being generalized. The coupling of recollections has already been illustrated in connexion with touch: when recollections of touch are combined with those of sight we have ideas of substantiality: weight and distance are clearly ideas in which recollections of objects are coupled with those of effort in lifting or approaching. Recollections are generalized when they are united with other recollections of the same kind: 'man' is a generalized idea formed by uniting all our recollections of individual men. These couplings and unifyings are made under the influence of certain attractive susceptibilities, which

[14] It has been concluded by some physiologists that we perceive our own nervous conditions by a special 'kinaesthetic' sense. But, if the vibrations of motor excitement can stimulate sensation, no such assumption is required.

may be generalized as the 'laws of thought'. It will be our task to elucidate them at a later stage.

When ideas connect themselves into a stream or current we have the reflective process known as 'thought'. Ideas pass in procession, like the pictures of a cinematograph, before our psychic selves, and stimulate physical and psychic responses. The idea-connexions [15] of thought are formed (we shall see) under the same laws as those which regulate the elaboration of recollections into concepts. The process need not be attended by motor excitement or feeling, since this is not inherent in ideas, but must come to them from outside. Every one has had experience of unconscious thought which may suddenly supply missing recollections and even form conclusions, by a hidden process, appropriately termed 'sub-conscious',[16] since it differs from reflex, or instinctive, nervous processes in involving the action of the brain. Thought becomes conscious when its ideas are illuminated by cerebral motor excitement. This is evidently derived from that which is generated in the process of sensation, for conscious thought is started by our fresh sensations on awaking. The first of them is attended by motor excitement which passes to the idea that is subsequently linked to it in thought; and since each idea-connexion emerges from obscurity and lapses into it again when its place is taken by another, we may conclude that the motor excitement flows along the chain of ideas, from link to link, like a current of electricity.

It may be urged that we are conscious of our dreams or sleeping-thoughts, at a time when sensation is suspended. In the condition of sleep the psychic and reflective planes are cut off from direct communication with the outside world, since they are isolated from the

[15] Called 'sentences' when formulated in words.

[16] Throughout these chapters the term 'sub-conscious' will be applied to nervous processes which are attended by 'unilluminated' ideas, whereas conscious processes are attended by 'illuminated' ideas, and unconscious processes by no ideas.

sense-organs.[17] Consequently psychic energy fails and the brain forms no sensations. But the physical nervous system can still be impressed; were this not so, no outside occurrence could awaken us. Accordingly, impressions, external or internal, may be received during sleep, and may find their way to the brain by an indirect channel, producing symbolic or imaginative sensations which develop motor excitement and become conscious as dreams. By dream-ideas psychic energy may be excited, as in somnambulism. But we remain asleep until direct communication is established between the psychic and cerebral planes and the world outside us.

If ideas can be thrown into groups, be coupled together, and can attract one another, it is difficult to avoid the conclusion that they possess physical substance—that they are specially modified brain cells. The hemispheres may, it seems, be figured as caskets containing myriads of little photographic plates upon which images are impressed by sensations and feelings. As experience widens, more and more plates are exposed. But plates remain for exposure. Our wonder is lessened at the strange insensibility of the brain tissue to pain—at the curious fact that animals, man included, may recover in time from the loss of considerable portions of it. Ferrier[18] writes that no 'reaction is produced by cutting, tearing, or even branding the surface. Human beings who have had their brains lacerated can testify to the entire absence of pain or suffering thereby'.

[17] Bastian cites a case of a youth of sixteen who had completely lost all conscious touch, all taste and smell, the sight of one eye, and the hearing of one ear. If these were closed 'after a few minutes the expression of surprise and the uneasy movements, which at first showed themselves, ceased, the respiration became regular, and in fact the patient was fast asleep' (*The Brain as an organ of the Mind*, p. 485).

[18] *Functions of the Brain*, p. 127. Flourens, quoted on page 125, is still more explicit: 'On peut retrancher, soit par devant, soit par derrière, soit par en haut, soit par côté, une partie assez étendue des lobes cérébraux sans que leurs fonctions soient perdues. Une portion assez restrainte de ces lobes suffit donc à l'exercice de leurs fonctions.'

The photographs would, however, be of characters, not of conscious ideas. To avoid ambiguity of thought we require a particular word which will distinguish the character which is possessed by a sensation from its illumination in consciousness. Let us extend to it the term 'idea'. If an idea is being formed in the process of sensation, it will be an *impression-idea* : if it has been formed and put away, it will be a *record-idea* : if it is elaborated in the brain from record-ideas, it will be a *concept-idea*. The word 'impression' will then signify the effect upon us of a stimulus, apart from any idea. These definite terms are the more required in that the word 'sensation' possesses an ambiguous signification : it denotes not only particular sensations, but the process of sensation. The latter cannot exist without the former and so we employ the same term for both. But if we wish to avoid indefiniteness of thought, the process should be distinguished from its fruits.

All conscious impression-ideas possess at least so much of the agreeable or disagreeable as renders them 'interesting', and by sympathetic interaction between their motor excitement and that which constitutes psychic energy they may become acutely pleasurable or displeasurable. Accordingly, it appears that the function of the brain is to enable us to profit by experience, instead of responding to stimuli automatically. For a pleasure or displeasure of present sensation is balanced through memory with the pleasure or displeasure of its *consequences,* and we experience this balancing in the process of choice. We appear, by an exercise of will, to resist the attraction of the greatest pleasure, or the repulsion of the greatest displeasure, that offers itself in this fashion. But willing, we shall see, is generally directed by ideas of its pleasurable consequences. For pleasure may be psychic as well as physical, and the pride of overcoming a temptation may be amongst our keenest delights. One may refuse a glass of wine to obtain the pleasure that is given by asserted dignity, as well as from apprehensions of a possible headache.

But alongside this life of choice there survives the life of instinct. Physiological experiment shows that it is far more influential than we suppose. Dogs have frequently survived for some months the loss of their brains. They are unconscious, but they retain their instinctive sensory susceptibilities and their powers of movement. Vertebrates of lower organization, such as frogs, exhibit the most extraordinary capacity for co-ordinated movement, even when decapitated. If a drop of acid be placed on the thigh of a frog, thus mutilated, it will endeavour to rub it off by bending up its foot, and, if the foot be held, it will make this attempt with the *foot of the other leg.* Ferrier's researches brought him to the conclusion [19] that 'it is an established fact that adapted actions, such as intelligence would also indicate, are capable of being called into play through our spinal cord entirely without consciousness'. So independent indeed is this unconscious activity that it has led physiologists to the abstract conception of the 'spinal animal'. In respect to his external movements man is, no doubt, less trammelled by instinct than the animals below him. But its influence remains. *Movements* evolve into *motives*—consequences into causes—the imperiousness of which we apostrophize as 'Fate'. We shall find, as we proceed, that even our choice of the most agreeable is a development of instinct. Its evolution can be traced. We can infer from the behaviour of all living creatures—from the humblest upwards—that it is instinctive to move towards the favourable and away from the unfavourable. Evolution has left this propensity unimpaired throughout the animal kingdom, but in conscious life has transformed it into a *motive* to seek the agreeable, since this is our idea of the favourable, and to avoid the disagreeable—the unfavourable in idea.

[19] *Functions of the Brain*, p. 20.

Moreover, this development of the conscious from the unconscious can be observed in the life of each individual. The earliest movements, gestures, and utterances of infancy are plainly instinctive. But ideas are formed of them, so that they are controlled, and may be initiated by the brain. Our conduct and speech are gradually elaborated by a process of 'individual' evolution, which leads each of us from the reflexes of infancy to the considered activities of mature age.

It will commonly be objected to our line of argument that it leads to conclusions which deprive life of all dignity and responsibility, exhibiting it as the consequence of external or internal stimuli. But there is a qualification which in a measure salves our pride. Our responses to stimuli depend very greatly upon our susceptibilities, and these may not only be modified, but be created by education. And, if it hurts our vanity to realize that we have little power over ourselves, we may be consoled by the thought that we have immense power over others. For we can suggest to them stimulating ideas which would otherwise have not influenced their feelings, thought, and conduct. Human society is a collection of units, each of which acts upon the units around it.

Before concluding this chapter, we may conveniently review the inferences that have been tentatively drawn :

1. An impression of the nerves generates a current of sensory excitement which is inevitably transformed into motor excitement. The sensory excitement may affect the physical, the psychic, or the reflective plane. In the first case the motor excitement, which proceeds from it, actuates a reflex movement ; in the second case it generates a continuous energy ; in the third case an enduring brain-stamp is formed of the character of the sensory excitement which the accompanying motor excitement illumines with agreeableness or disagreeableness—that is to say, renders conscious. Accordingly, motor excitement may be muscular, psychic, or cerebral.

2. Sensations may be of our own muscular, neural, or psychic conditions as well as of things outside us.

3. A recollection is a record of the character of a sensation. Such records are coupled and generalized to form concepts, and these are strung into chains of thought by the action of certain definite susceptibilities. During conscious thought a current of cerebral motor excitement, originally derived from sensation, runs along the chain, from link to link.

4. Since consciousness is pleasure or displeasure, its effect is to reduce different impressions and recollections to a common denominator, and so render them comparable. This enables us to profit by experience, since on a comparison of pleasures and displeasures we are swayed, respectively, by the greatest and the least.

5. Consequently pleasures and displeasures may compete with instinctive impulses as motives of conduct. They become active motives by being associated with the physical, or instinctive, propensities to approach the favourable, and recoil from the unfavourable.

CHAPTER II

THE PHYSICAL PLANE

THE conversion of sensory excitement into motor excitement obviously involves a change of condition in the nerve-cell, and, unless this change occurs, a stimulus does not affect us. An injurious touch, for example, stimulates a movement of recoil; but, before this takes place, a nervous change occurs which produces a condition of repulsive excitement. In this case the external impression suffices of itself to bring about the nervous change. Of this class are the stimuli that are ordinarily called 'sensory', such as touches and familiar sights and sounds. In other cases the effect of an external stimulus depends upon the association with it of a contributory nervous state. For instance, an incident which alarms us when it is strange may leave us unmoved when it has become familiar: a difficulty does not confuse us unless we are making an effort which it obstructs. We may distinguish stimuli of this class as 'neuro-sensory'. In other cases nerves are stimulated by changes in the conditions of body tissue, or by glandular secretions. An appetite—hunger, for example—results from certain bodily conditions which press us to exertion quite independently of any external sensory impressions: peculiar secretions (*hormones*)[1] are passed into the blood on particular occasions and stimulate into temporary activity nerves that initiate certain periodic bodily processes—those of digestion, for instance. Stimuli of this class may be termed 'organic'. Nerves may also be stimulated by changes that occur in the

[1] Alcohol, drugs, and many medicines may perhaps be regarded as artificial *hormones*. And microbes *may* be harmful nerve-stimuli.

nervous system itself and lead to nervous consequences: a spasm of fear may stimulate a revulsion of relief in nervous succession—'neuro-serially', as we may style it. A series of such organic and neuro-serial stimulations appears to activate the automatic co-ordinated activities which we ordinarily characterize as 'instinctive'. They culminate in the functioning of our internal organs which involve processes that are more complicated than the most amazing dexterities of insect life —and are as mysterious in their guiding purpose.

In discussing the effect of stimuli upon the physical, or instinctive, nervous system, it will be convenient to treat of the organic, neuro-sensory, and neuro-serial classes before the sensory. It is not that one is more essential than the other: an animal would perish if it experienced no hunger, but it would also perish were it not stimulated by impressions of its food. But since a certain internal condition must precede an instinctive response, whereas an external impression may or may not precede it, internal conditions are of a more primitive character than impressions. Before, however, coming to this subject we may insist again upon the fact that instinctive stimuli affect us independently of conscious sensation. We may be conscious of them, but our ideas of them are ancillary to their action, not essential. We are conscious of the conditions that stimulate the appetites—conditions which we feel as hunger, thirst, lust, and fatigue. But certain of them may affect us during sleep. And they are evidently experienced unconsciously by animals whose nervous organization does not rise above the physical plane. It has, moreover, been proved that vertebrate animals whose brain has been excised respond, nevertheless, to instinctive stimuli—a fact which will be illustrated as we proceed. The feelings and ideas which may accompany instinctive neural states are, then, merely ancillary to them; and if we would isolate such of our nervous experiences as are essentially physical, we must distinguish between sensations or feelings which, although conscious, would affect us independently of consciousness, and those which affect us by being conscious. This is only possible by means of inferences drawn from our unconscious actions, and from similarities between the anatomy and behaviour of brainless animals and our own.

First, then, of the stimuli which we have classed as *organic.* We are moved by them in our appetites—imperious promptings to eat, drink, procreate, sleep, and evacuate, that are stimulated by the unfavourable conditions of which we are conscious as hunger, thirst, lust, fatigue, and the discomfort of congestion. In so far as these experiences are actually painful they probably affect the brain through impressions of internal touch. It has been proved that the pain of hunger results from a constricted [2] or cramped condition of the stomach. The disquieting sensations which the appetites occasion bring them within the scope of reflective life: if their promptings were unattended by ideas, they would be altogether beyond the control of the brain. Even as it is, the appetites are masterful impulses: when urgently pressing, they yield to fear, but to no other impulse. In the extremity of famine mothers will sell their children for a few pence; tormented by lust a man will kill the woman he desires, and we may find in its irritating influence an explanation of many sexual aberrations.

It has been proved [3] by physiological experiment that the effect of the appetites is unconscious as well as conscious. Dogs which have been deprived of their brains grow restless at feeding time, and are periodically affected by sleep and wakefulness; a male pigeon, after excision of its brain, shows distinct traces of sexual excitement, and in spring will coo all day long.

The appetites stimulate reflex movements appropriate

[2] Cannon's *Bodily Changes in Pain, Hunger, Fear, and Anger* (Appleton, New York).
[3] Schafer's *Text-book of Physiology*, ii. 998.

to their necessities. Such are the grippings, suckings, and bitings of infancy, and the lying-down which is the instinctive response to fatigue. We form ideas of them, and can control and direct them. But swallowing remains a reflex throughout life.

When, under the influence of an appetite, we receive an impression of something that satisfies it, the neural stimulus becomes *neuro-sensory*, and a condition of irritation gives place to one of attraction. So we are drawn to our food, and there arise the compelling fondnesses of sexual and maternal love. The appeasement of an appetite is followed by a *neuro-serial* state of restfulness which checks further movement. So long as it is not allayed, a condition of restlessness continues which prompts continued effort. We are conscious of these states as satisfaction and dissatisfaction.

Stimuli of the neuro-sensory class are strikingly illustrated by the effect of the *strange*. This gives a shock to the nerves that is followed by a neural state of disturbance, and the active movements of attention or escape. We attribute these experiences to the sensory impression that precedes them in consciousness. But they are clearly due, not to the character of the impression, but to its strangeness, for, when it has become familiar it may cease to excite them. It causes a disturbance of nervous equilibrium because it is out of accord, or relationship, with past experience. This implies the existence of a primitive form of memory in the nerve-cells—a conclusion which will seem less surprising as we proceed. If memory has *evolved*, we must expect to find its rudiments far down the scale of life. We can satisfy ourselves by the behaviour of lowly organized creatures that the shock caused by the strange is independent of consciousness : when suddenly confronted by strange impressions, the humblest of them exhibit movements of repulsion. And, in ourselves, attention clearly must precede consciousness, and the concentration of susceptibility which it involves must then be an instinctive[4] effort. Attention is arrested most forcibly by the sudden environal changes that arouse us from sleep—by touches, glaring light, or strident sounds. But any distinct change in our environment will suffice—any impression, internal or external, that is out of the ordinary—a fact of which the art of advertisement takes full advantage. The startling effect of the impression is increased if it is in very strong contrast to present experience: it is then followed by the involuntary start of which we are conscious as alarm.

The familiar, on the other hand, leaves the nerves in a condition of peaceful equilibrium which is generally favourable to life. Its influence is consequently attractive—far more attractive than we consciously realize. It affects us physically: the stomach rejects strange food, but will accept any non-poisonous substance to which it has grown accustomed.[5] We may reasonably assume that the instinctive attraction of the familiar is the ultimate basis of gregarious life, for this is developed as strongly amongst brainless creatures as amongst the higher animals. We are conscious of its agreeableness in the charm of 'old familiar faces', in the pleasure with which we revisit the scenes of childhood, and in our preference of acquaintances to strangers. But its hold upon us is masked by a rival susceptibility—the attractiveness of the new, that is to say, the impulse of curiosity. This impels us to enlarge our experience by connecting new impressions with those that have

[4] 'Hardly has the object thrown its image on the retina than the ocular muscles place the eye in an appropriate position for receiving the image, the movements of the two eyes being mutually co-ordinated. Other reflex movements take place, regulating the diameter of the diaphragm formed by the iris, and also the curvature of the crystalline lens' (Morat's *Physiology of the Nervous System*, p. 577). 'In the ear the muscles of the ossicles tighten or relax the tympanum' (p. 627). 'In the movements of the nostrils and those of the thorax we observe (in cases of attention to smells) almost exactly the action of the directing muscles of the eye and the external ear' (p. 648).

[5] We are reminded of the advice of a celebrated physician to one who, having been recommended to take a little beer, objected that it disagreed with him. 'Drink it until it doesn't' was his reply.

already been formed: it is a psychic reaction and involves more elaborate nervous processes than those in the physical plane. If we doubt whether the influence of the familiar is as strong as is now inferred, we may be reassured by one of the most striking facts in human nature—the extraordinary hold of habit. The courses of action to which we habitually cleave—from which it is so difficult to stray—are suggested by memory. But memory, of itself, is not a motive: there must be an inclination to accept its suggestions, and this, we may be sure, comes from their familiarity. And familiarity obviously contributes to the bond which attaches us to our 'favourite' likes and dislikes. It is by practice—by forcing familiarity—that one 'learns to smoke', through, it may be, very painful experiences.

There is a similar contrast between the nervous effect of the injurious, the difficult, and the antagonistic, and of the beneficial, the easy, and the helpful. The injurious causes the alarm which becomes the emotion of fear when it is translated into psychic excitement: it is followed by a *movement of escape*. The difficult or antagonistic may be experienced in approaching the favourable or in recoiling from the unfavourable, or in the appeasement of an appetite. It causes the confusion which is felt as dread: a neuro-serial impulse follows in an *effort of venture*. This may be random and spasmodic, as when we frantically endeavour to open in haste a railway carriage door, or try to fit a key to a lock in the dark. But, guided by ideas, it becomes the calculated experimenting that is stimulated by problems.

The behaviour of brainless animals will convince us that impulsive endeavours which we associate with consciousness are in their primitive phases unconscious. Animalcules that are prevented by an obstacle from rising to the top of a glass of water will by repeated ventures find a way round it, whereas air-bubbles simply adhere to its under-surface.

When excited favourably or unfavourably the nerves are thrown out of equilibrium. But they tend to recover their balance. If the strange is not really alarming, or if a danger or difficulty is averted, there is the nervous recovery of which we are conscious as *relief.* This is a response of the neuro-serial class: it is a nervous reaction. If an endeavour to satisfy an appetite, a struggle of escape, or an effort of venture achieves itself, there ensues the favourable neural state of which we are conscious as *success*: if it does not achieve itself, it is followed by the neural state which we call *failure.* These are, again, neuro-serial responses. Success and failure are plainly ideas, not of certain acts or methods of conduct, but of their nervous *results.* They are neural states that may be figured to be of expansion and contraction, which have an active effect in checking further movement or in prolonging it. A butterfly that succeeds in finding honey in a flower comes to rest; if it fails, it pursues its flight elsewhere. Success and failure involve sharp nervous reactions before the nerves settle down into equilibrium. These, however, can be discussed most appropriately in the chapter following.

We must include amongst neuro-sensory stimuli the effect of rhythm. It gives a pleasure which is common to all mankind: it is appreciated by some of the birds,[*] and appears to touch some insects. We commonly regard it as a sensation of hearing. But we can feel its influence through touch, as in dancing, and through sight, as in the agreeable effect of the rhythmic repetition of a decorative design. It gives a peculiar charm to poetry. Its stimulating effect seems, then, to have a deeper foundation than the sense organs, and to arise from a peculiar physical susceptibility. For rhythm is

[*] Peacocks and cranes, for instance, dance with rhythmic movements. And no one who has watched myriads of fire-flies flashing and darkening their light simultaneously, will deny that these insects are susceptible to rhythm.

in harmony with the instinctive life of all animals whose circulation is effected by rhythmic pulsation. The beating of the heart is maintained by an endless chain of neural stimuli—by a nervous susceptibility to rhythmic changes. It sets the body into a continuous state of rhythm of which we are conscious at moments when our attention is not arrested by other stimuli. It is then that we become conscious of the passage of time—that time, as it is said, 'hangs heavy upon our hands'. For time, as a duration, is a continuing rhythm—a fact which is acknowledged in our use of the same word to express time and rhythm. It may, then, well be that the physical nervous system responds sympathetically to external impressions of rhythm. The drum resounds throughout the world: dancing is a pastime with the rudest of savages.

Finally, of *sensory* stimuli which impress us through the sense-organs. The most archaic of the senses is obviously that of touch: it is possessed by microscopic creatures that lack all vestiges of specialized sense-organs. It is clearly the effect of pressure upon the nerves, which, if acutely unfavourable, gives us the sensation of physical pain, and excites a movement of recoil. The nerves can be pressed by congested body tissue, and accordingly pain may be internal. Touches that are favourable to the life of the individual, or the species, stimulate approach, and are in consciousness acutely pleasurable. Consciousness apart, however, touch enables an animal to distinguish between the favourable and the unfavourable, as for instance in regard to food, provided that its nerves are susceptible to one and the other. And, through changes in the pressure of body-tissue upon the nerves, touch renders it susceptible to differences and changes of temperature. For heat and cold are accompanied by expansion and contraction of tissue, which would vary its tension.

The evolution of the sense of smell afforded an animal its primitive means of detecting objects at a distance. If it is in active movement, smell directs it through increasing or diminishing intensity. In vertebrate animals smell is believed to be the result of the chemical action of minute particles of matter. But in their case the sense of smell is developed in a peculiar fashion: the olfactory nerve is in fact the embryo of the brain. Amongst invertebrates smell may be a form of touch: changes in tissue pressure would result from the exciting or irritating effect of particles of matter, such as are grossly illustrated by the effect of a mustard plaster. Taste is a combination [7] of touch and smell, a fact which is appreciated by every one who has held his nose when taking medicine.

By the senses of sight and hearing the detection and discrimination of objects at a distance is extended immensely. We may reasonably regard them as elaborate developments of the sense of touch.[8] For the vibrations of light and the air-waves of sound cause pressure. Minute creatures that are devoid of any specialized sense organs may be sensitive to sound and light—indeed, some of them are affected by the actinic rays of light which are beyond the susceptibilities of the human eye.

There is much difficulty in determining what sensory impressions move us instinctively, since a vastly large proportion of our sensations appeal to special susceptibilities—or 'tastes'—which are created by a process of conscious *individual* evolution, that is to say, are evolved through the action of the brain by a process which begins at each man's birth and continues through his lifetime. They stand, therefore, upon a higher level than the physical. Tastes in food, colour, and sound differ in an extraordinary fashion from nation to nation,

[7] 'We project our olfactory sensations into our mouth' (Professor J. B. Haycraft in Schafer's *Text-book of Physiology*, ii. 1258).

[8] An interesting development of the sense of touch is the function of the semi-circular canals of the ear. These may be likened to three spirit-levels, set at different angles, which inform us of the position of the head (Dr. Isaac H. Jones, *Equilibrium and Vertigo*).

and have varied continually during the past. These complexities of susceptibility are the fundamental differences between savagery and civilization. Impressions that affect us instinctively must be common to all races of mankind, cultured and uncultured, and it is a remarkable fact that—sensations of touch apart—very few sensory likes and dislikes are universal. Speaking generally, the pleasures and displeasures that come from the sense of touch are the same for all men. Perhaps the only exception is the pleasure of exercise for its own sake: this is not appreciated by the savage. Sensations of smell and taste have become vastly complicated, and in their case mankind is uniform only in disliking the pungent, acid and bitter, and in liking food and drink of some kind, the sweet and aromatic. For the rest, 'one man's meat is another man's poison'. So again with sensations of sight and hearing. All men naturally like brilliant colour and light which is not glaring; all men are naturally pleased by sound which is not harsh or strident. But there are the most astonishing differences between one race and another in respect to the evolved pleasures that are called 'aesthetic'.

It seems, then, that the sensory impressions which affect us instinctively are those of touch, and the crudest of those that reach us through smell, sight, and hearing. These must stimulate us unconsciously as well as consciously, for they can awaken us from sleep.[9] And it has been [10] demonstrated that dogs whose brains have been excised will shut their eyes against glaring light, can be aroused from sleep by a strident sound, will reject food that is flavoured with quinine, and will sneeze in response to tobacco smoke.

Muscular activity of some kind is the instinctive outcome of a neural change: such are the movements by

[9] Changes of temperature affect us when asleep. We may throw off the bedclothes, and assume positions of which we are quite unaware until we awake.

[10] Schafer's *Text-book of Physiology*, ii. 702. See also Luciani's *Human Physiology*, iii. 509.

which the appetites are satisfied, the arrest of attention, starts of alarm, efforts of venture, and the approach or recoil that is stimulated by favourable or unfavourable sensory impressions. The most elementary of them persist as reflexes. But the ideas that are formed of them bring them under voluntary control: we can cough or wink voluntarily, and with such variations as we please; we can deliberately concentrate our attention. As *movements*, reflexes have contributed but little to the evolution of our methods of conduct. But they powerfully influence our conscious life as *motives*; and one of them—the effort of venture—under the influence of psychic energy and the direction of the brain, evolves into the Will.

The gist of this chapter is that the most fundamental of our susceptibilities and our impulses affect us[11] unconsciously as well as consciously, so that we are unaware of the real extent of their power. The stimuli that move us may be external sensory impressions. But they may also arise from organic conditions, from neural reactions, or from sensory impressions and neural reactions in combination. Amongst sensory impressions those of touch are markedly instinctive. And we owe to instinct our dislike of the strange, and liking for the familiar, the beginnings of fear and dread, our responsiveness to rhythm, and our appetites. The neural states of attraction and repulsion, of restfulness and restlessness, of exaltation and depression, of satisfaction and dissatisfaction are also instinctive; but they are the foundations from which have evolved the psychic states of which we are conscious as pleasure and displeasure and the emotions.

[11] The fact that response is more primitive than consciousness is the starting-point for the evolution of *associative stimulation*. An idea that has been associated with a response, muscular or emotional, stimulates it by the force of its association.

CHAPTER III

PSYCHIC ENERGY

Were not surprise blunted by habit, we should marvel at the oft-recurring changes in the amount of our energy or 'psychic force'. It appears to shrink into nothingness during sleep—just as the electricity of an electrical machine dies away when the handle is no longer turned. In our waking hours it varies from a high-strung pitch that renders us alert, quick in decision, 'ready for anything'—to use a colloquialism—to a condition of slackness in which we are listless, bored and undecided. These conditions may be more or less chronic in the individual. Some men are characteristically energetic: others apathetic. We commonly ascribe the difference to good or bad health. It is favourable to the development of energy that the organs of the body should function properly, that food and sleep should be sufficient. But these conditions are not essential. Some of the most forceful men have been invalids: Wolfe is a conspicuous instance. They possess an elasticity of spirit which prevails over bodily weakness. Others, with every physical condition in their favour, are invincibly apathetic.

This energy is not identical with life. We do not die when it disappears, as during sleep or in a fainting fit. But it seems to be as mysterious as life. We individualize it as 'the soul'. We also term it 'the spirit', identifying it with the breath of life. But breathing continues during sleep, when this energy deserts us. There is a painful air of sacrilege in attempting to trace it to an evolutionary origin. Yet it is clearly not peculiar to man. Dogs display good or bad spirits, are energetic or lazy, and can resist temptation. We can assign it to a nervous foundation if we assume that nerve-cells, when massed into ganglia, can be maintained in a state of continuous excitement by sensory impressions that generate motor excitement which is not discharged in muscular movement. This assumption is supported by the facts (1) that psychic energy revives with our first impressions on awaking; (2) that impressions, as we have seen, may stimulate the spinal cord directly; and (3) that sensory excitement is certainly transformed into motor excitement in the spinal cord, and that this excitement, if not discharged in muscular movement, must find an outlet. Such an outlet would be provided were it employed in setting a mass of nerve-cells into continuous vibration, and so producing a condition which may be figured as one of elastic nervous tension.[1] From the fact that psychic energy is hardly affected by blindness or deafness we can infer that the stimuli which set up this condition are those of touch, internal and external, and of exciting neural conditions on the physical plane, the effect of which during sleep is cut off from the spinal cord.

We can argue from our feelings that this continuing energy is exceedingly sensitive to the motor excitement which is generated in conscious sensation and is transmitted during thought, being thrown by it into states of sympathetic vibratory activity which are felt in consciousness as pleasurable or displeasurable. We may also infer that it is elastic—that having been thrown into one of these states it tends to spring back from it. It is accordingly not merely a reserve of energy, but through its elasticity can stimulate fresh nervous activity. It may be said for this hypothesis that it explains the extraordinary fluctuations of mood that are caused by impressions and ideas.

[1] Physiology assumes the existence of a 'continued glow of excitement in the spinal motor nerves' as a source of the enduring tension of the body muscles during waking hours.

That motor excitement—in consciousness psychic *feeling*—is energy is clear from the fact that it needs a safety valve. When we are 'charged' with excitement we use muscular movements of no immediate value as an outlet for an overflow. This is obviously the primitive function of our 'expressions of emotions' —gesticulations and ejaculations, frowns and tears, smiles and laughter. We can feel that they relieve the pressure of excitement, which, if they are repressed, may become intolerable, or produce the collapse of a fainting fit.

It is also clear that nervous excitement may be of two very different kinds, presented in feeling as agreeable and disagreeable. We are pleasantly excited, or exhilarated, by agreeable impressions—as by bracing weather, by the taste of our food or by music. We are also excited when depressed, in the nervous conditions produced by unfavourable or disagreeable impressions or ideas. We do not realize that there is energy in these conditions because it is, so to speak, 'contractive' instead of 'expansive'; it causes us to shrink away from the object or cause of sensation, not to expand towards it—a feeling, which is experienced very strongly in physical pain, in fear and in the agony of sorrow and shame. That these conditions involve energy is, however, evident from the muscular and glandular activity which they provoke; grief manifests itself by violent gesticulations, by contraction of the facial muscles, by sobs and tears, just as joy does by expansive movements. It is, indeed, so exhausting that it rapidly brings on the apathy of fatigue.

Motor excitement must accordingly be of two kinds or phases. We have conjectured that it is vibratory on the ground that, since its primitive function is to stimulate movement, it is probably of a mechanical nature. Taken by itself this reason is far from convincing. But the theory is greatly strengthened by the consideration that, if the excitement were vibratory, it would, by stimulating the sensory nerves, provide an origin for the ideas of psychic excitement which, combined with the excitement, are our *feelings*. On this hypothesis the two phases of motor excitement are two phases of vibration. It may be that these differ in being accordant or discordant—a supposition which is borne out by the emotional effects of concords and discords in music, but remains highly conjectural. We are, however, on firmer ground if we attempt to distinguish them by their muscular effects. It is characteristic of living tissue that it expands under favourable, contracts under unfavourable, circumstances. These responses are exhibited very clearly by the unicellular microscopic creatures which may be regarded as single nerve-cells living independently; they are also noticeable in animals of higher class that are soft-bodied, such as zoophytes, worms, and molluscs. These all expand under favourable conditions of food, light, and temperature, contract, if circumstances are adverse, or they receive strange or injurious touches. And, turning to our own experiences, the terms 'expansive' and 'contractive' aptly describe the feelings which move us when exhilarated or depressed. We may, then, conveniently use these terms to distinguish the two phases of vibratory motor excitement, stimulated, respectively, by the favourable and the unfavourable, remembering that they express the *effect* of these nervous conditions upon the body tissues.

We cannot but be curious as to the precise part of the nervous system which by its vibrations provides us with psychic energy or the spirit of emotion. The vibrating cells may form part of the spinal cord, which includes complexities of nervous tissue the functions of which are still undetermined. But it would certainly appear probable that, in the course of evolution, a special nerve system would have become appropriated to this purpose. Such appears to be the cerebellum. The use of this large ganglion is an unsettled problem. It is connected with the spinal cord on the one hand, and with the brain on the other, offering an alternative route for impulses passing between them. It can, accordingly, be actuated by ideas of the brain, and it can influence the spinal cord by its activity. We are, it will be objected, indulging in purely imaginative speculation. Not so: this hypothesis is supported by much that has been discovered by the physiological experiments of the last half century, and by the conclusions that have been formed by some eminent authorities.[2] It is admittedly a theory. But we are peering through the veil which conceals one of Nature's profoundest mysteries, and our only hope of understanding such facts as we can dimly perceive is to bring them together by an inferential hypothesis which will at once guide further observation and be brought to test by it.

Our sensations are only in part derived from the world around us: they are in very great measure conscious impressions of ourselves. We term these impressions 'feelings'. The internal bodily or nervous conditions, which we *feel*, generally include both physical and psychic elements, but may be wholly psychic, as in those of pride and amusement. We must sharply distinguish the *conditions* from the *feelings*, or sensations, in which they become conscious. Nervous conditions and changes are *facts* which occur in the lives of brainless animals, as well as in the lives of animals with brains. Sensations of these facts are their reflections in the brain. Accordingly two elements enter into a feeling—a physical or nervous fact, or state, and an impression-idea, or sensation of it. In conscious life the facts and their reflections occur together, and it is by no means easy to distinguish them apart—to analyse a feeling into a state

[2] Summarized in Note A appended to this chapter. It may be observed that the cerebellum is highly developed in birds. They are excessively energetic, and their song is one of the most striking expressions of psychic excitement.

and a sensation of the state, and to realize that fear is the consciousness of a state of nervous perturbation which exists independently of consciousness. The idea of the state that is formed by the brain may exist apart from the state. But the two are confused in language. 'Courage', for instance, means a nervous state of courage, and also an idea of this state.

That a conscious feeling has no existence apart from the brain seems to be proved by physiological observation and experiment. It may be, however, that in psychic motor excitement there is just so much admixture of sensory excitement as would produce a vague consciousness; and it is to be said for this surmise that it would establish a stepping-stone for the evolution of the brain. It is difficult to watch a butterfly expanding itself in the sunlight, and shrinking in bad weather underneath a leaf, without believing that it *feels* some degree of exhilaration or depression. But its feelings would be idealess, and vague in the extreme.

Our psychic energy is the material from which originate the 'spiritual' element of the impressions called *feelings*, and of the ideas that are derived from them. It plays another part of immense importance. All conscious impressions, as we have seen, are agreeable or disagreeable, according to the phase of the cerebral motor excitement which enters into them. But their character as such is immensely intensified, and becomes acutely pleasurable or displeasurable, when the cerebral motor excitement is reinforced[3] by that of the psychic plane. This interaction between the psychic and cerebral planes will be illustrated as we proceed. Its actuality is demonstrated by the fact that a wound or a burn will hardly be felt by one who is in a condition of strong expansive excitement, since in this case the effect of the injury upon the brain is not reinforced but

[3] One can feel the pain of a sudden injury grow from discomfort to agony as cerebral and psychic motor excitement mutually stimulate one another.

counteracted. We can comprehend without difficulty how the interaction comes about. For cerebral and psychic motor excitement are alike in being vibratory, and can affect each other sympathetically after the fashion that the notes of one musical instrument arouse sounds in another. It may be observed that the psychic excitement which sympathetically affects the brain in this manner can be aroused by an impression of the brain, as well as by a physical stimulus, whether organic, neuro-sensory, or sensory. Thus we are exhilarated or depressed by sights of the beautiful or the miserable, as well as by hunger, satisfaction, fear, or touches. In this case the cerebral motor excitement which attends the impression stimulates motor excitement on the psychic plane, and is itself reinforced by it.

We must, then, carefully distinguish between the effect of psychic excitement in giving rise to impression-ideas of internal energies, and of their intensities, and its influence upon the pleasurable or displeasurable motor excitement of the brain. A fit of anger, for instance, impresses the brain as an idea of strong discordant energy, associated with the effects of a stimulus. At the same time it sympathetically reinforces the unpleasantness of this idea, and may, indeed, infect with displeasure all accompanying impressions, such as those of the furniture by which we are surrounded at the time.

The effect of psychic conditions in infecting our impressions with pleasure or displeasure is strikingly illustrated by our moods. These are tonic states of good or bad spirits—of happiness or unhappiness—which may, perhaps, be dimly felt independently of the brain, but assert themselves in consciousness by colouring the impressions, feelings, and thoughts of the moment. They are influenced by ideas of the brain, as by visions of success or failure, as well as by purely physical causes—by food and rest, the state of our health, the temperature and the weather. We are 'braced' by dry or keen air. We may also brace ourselves by artificial means. The neural effects of alcohol generate expansive excitement, and throughout the world human ingenuity has set itself to procure this means of happiness. Its excitement is, however, unstable, and may be reversed by ideas, so that intoxication may be maudlin or quarrelsome as well as cheerful. The exhilaration of opium and the resin of the hemp plant is less inconstant. But the effect of alcohol and narcotics is followed by a grievous revulsion when the stimulus passes off. The influence of tea and tobacco is less marked; but they undoubtedly enable us to withstand irritation, and to take the detached view of life which is termed 'philosophic'. It accords with our theory of the vibratory nature of psychic excitement that it is peculiarly sensitive to rhythm, whether in sound or in bodily movement. Music affects it acutely, and is employed as a stimulant in religion and in war. Dancing invigorates like champagne, and is very much more wholesome.

Moods may simply energize nervous conditions that exist on the physical plane. Exhilaration and depression may be the intensified counterparts of favourable or unfavourable physical conditions. Accordingly we may be exhilarated or depressed in spirits by a meal or an injury. But these psychic conditions, once established, may in their turn modify our physical susceptibilities. In good spirits we are charmed by impressions, such as those of the beauties of Nature which in more sober moments might leave us unmoved: in bad spirits petty crosses may become too heavy to be borne. Moreover, psychic energy, being a distinct evolutionary development, can maintain itself in some degree independently of physical conditions, and the effect of physical stimuli; and can even expansively resist, or contractively overflood them. Of this momentous fact our everyday experiences assure us. Psychic exhilaration may render annoyances amusing: psychic depression may blight our appreciation of the pleasing.

Resistance may be complete or incomplete. It may overcome depression altogether, as courage annihilates fear; or it may infuse physical depression with expansive activity, such as converts depression into irritation. And, since psychic energy is stimulated by ideas, it becomes a medium by which the mental can antagonize and overcome the physical.

A mood is a *passive* condition. Psychic excitement is in itself passive; it becomes active when it is influenced by an instinctive impulse, but, even then, it remains a 'passion'. But we must not press this point too far. For psychic excitement is certainly active when it *resists* the effect of stimuli. Moreover, since nervous excitement liberates itself in muscular movement, the psychic states which we call 'passive' translate themselves into muscular activities that are peculiar to them. When normally expansive they free themselves in bracing the muscles: when stimulated, or reversed in phase, they are relieved by the multiform movements and secretions that are called 'expressions of emotion'. They also find an escape in the brain, and give thought an emotional complexion or turn it into *expressive*, or imaginative, channels.

The elasticity of psychic energy is strikingly illustrated by the revulsion of feeling which occurs when a stimulus is suddenly withdrawn. When we are reassured by finding that what seemed strange or injurious is not actually so, we 'smile at our fears'. Joy and sorrow are revulsions, and are accordingly spiritual crises. Pleasure becomes joy and displeasure grief, when the effect of a stimulus is reinforced by a change of mood; pleasure becomes grief and displeasure joy, when a stimulus is withdrawn, as if a loved one is removed by parting or death, or we are suddenly relieved from pain. Revulsions are an abounding source of sorrow. Yet we can draw from their consequences an optimistic conclusion, which we feel to be true—that, fortunately for mankind, expansive conditions are the normal. Reversed, even by physical pain and shame,[4] they are more vivacious than contractive excitement, and strive to recover themselves. No one will dispute the truth of the saying that 'it is better to have loved and lost than never to have loved at all'. A revulsion from a trifling displeasure is acutely pleasurable: it is as if a pendulum, drawn an inch across, swung back a foot when released. The importance of this fact cannot be overrated. It is the source of the pleasure which we draw from the ludicrous, and from our amusements.

Resistance is, however, the most striking manifestation of psychic energy, since it introduces a new impulse into life—that of active antagonism, which is altogether lacking in creatures whose nerves are unconcentrated. They are merely capable of efforts of escape or random venture. But when a mood resists the physical effects of the difficult, the strange or the antagonistic, the nervous condition of confusion is steadied, fear becomes courage, depression becomes irritation which is emotionalized as anger. And we shall come to the momentous conclusion that psychic resistance can stimulate the effort which is the physical response to nervous confusion, and, in conjunction with it, evolves into the Will.

It may be observed here that, by spiritualizing the *consequences* of effort, psychic energy evolves the feelings of respect and pity from reactions which follow upon success and failure. These are nervous physical conditions of violent inequilibrium from which recovery is made by an automatic 'shrinking' or 'swelling'—that is to say, by a contractive or expansive reaction—which when spiritualized and associated with the personality that is the *cause* of the success or failure, becomes in consciousness the feeling of respect or pity. We cannot help respecting ourselves, if successful, or pitying ourselves in case of failure or distress; and, if ideas of

[4] Shame is so grievous because the depression of failure is aggravated by the self-contempt which it excites.

success or failure have become associated with another, respect [5] or pity will be aroused whenever an idea of him occurs to us. Moreover, since power, superiority, or excellence are traits of success, and weakness, inferiority, or 'misery' traits of failure, ideas of these traits, when associated consequentially with ourselves or another, will suffice to excite feelings of respect or pity.

It follows that the neural states of success and failure are the primitive *causes*, or stimuli, of the reactions of respect and pity; they are, in fact, *neuro-serial* stimuli. A similar neuro-serial reaction follows the relief that comes from the intervention of the protective power of another. Associated with him this becomes emotional as the psychic *feeling* of confidence or faith. So close is the connexion of these neural causes and consequences that the *idea* of one will *actually* stimulate the other in reverse order. Manners that imply respect for ourselves, or complimentary words, arouse in us a feeling of superiority. The idea of another's pity subjects us to a feeling [6] of misery. We therefore dislike to be pitied. And if we have faith in another, we feel convincingly his protective power.

The instinctive neural states of success and failure, *in themselves*, are spiritualized by the energetic psychic revulsions of which we are conscious as pride [7] and shame. Both are revulsions from the strain of resistance—pride an expansion that follows the relaxation of success, shame a contraction that follows the relaxation of failure. Pride is, then, a form of joy, shame a form of sorrow. Yet they differ from joy or

[5] Under the action of the brain respect may be reversed into jealousy, and pity into scorn. But we must postpone consideration of these metamorphoses to a later chapter.

[6] Every mother knows that a child cries all the more if pitied when crying.

[7] Throughout this book the word 'pride' is used in its original, not in its derivative sense. It signifies the triumphant exhilaration that follows a successful effort, not the continued state of self-complacency that is maintained by the association of ideas of success with oneself.

sorrow, ordinarily so called. For they result not from the effect of a stimulus, or of its withdrawal, but from successfully or unsuccessfully resisting its effect. They appear to be caused by ourselves, independently of our environment. Accordingly pride, as the consequence of resistance, is an inspiriting feeling of successful independence which stimulates our egotism or self-love. We love ourselves in success, hate ourselves when ashamed. It is the origin of our 'dignity'—a concept in which ideas of pride and respect are combined in sequence. If the sequence be reversed, the idea is of our 'honour'. One who had lost all sense of dignity or honour would realize that he is the sport of circumstances, the plaything of Fate. The successful resistance on which we pride ourselves is, nevertheless, the consequence of a stimulus. But, since it runs counter to the stimulus, it appears to be spontaneous.

Resistance is opposition to the incompatible. Under the influence of ideas it develops into the various phases of effort which are termed 'willing'. But it may be simply the consequence of psychic tension, acting independently of the brain. Displeasure, and the confusion of fear and dread, are incompatible with psychic expansive excitement: pleasure with displeasure: failure with pride. Accordingly fear, dread, and failure are resisted in courage, anger, emulation, and industry. Following out this idea we can infer—and feel—that there is incompatibility between the physical and the psychic, and between the automatic sequences of stimulated response (and of habit) and self-conscious egotism. This manifests itself in resistance to physical stimuli—an antagonism in which we shall find the ultimate source of asceticism and morality; and we can feel that the 'assertive' volition termed 'spontaneous' is a protest against the automatic. Resistance is *practical* because it stimulates and develops a physical element—the effort which is the instinctive response to alarm and confusion. This may take the form of contrariety. So a naughty child may refuse its food; and we term its behaviour 'wilful', recognizing that in mere petulance there are the rudiments of willing. It is an effort of venture in assertive volition, and when it takes the form of experiment. In the willing of *choice*, it becomes an impulse of pursuit or avoidance.

When a stimulus is not resisted, its nervous effect is immensely intensified by the emotional, or spiritual, excitement which it produces on the psychic plane. Alarm is emotionalized into fear, love into a passion. Conditions of satisfaction and relief become joys which, when associated with the person who causes them, become the attachment of love or affection. Dissatisfaction and unrest become griefs which are similarly associated as hate. The peacefulness of the familiar, in itself, produces little psychic effect, and for this reason we are unaware of its real hold upon us. But, as a revulsion from the strain of war, it sets the world dancing.

We must refer again in more detail to the function of psychic excitement in sympathetically enhancing the agreeable and disagreeable into pleasure and displeasure. The association of pleasure or displeasure in different degrees with the sensory impressions of all our experiences has the effect of 'correlating' them—of reducing stimuli to two common denominators, as involving either pleasure or displeasure of different intensities, so that stimuli of various kinds can be compared with one another in expectation. Since the pleasurable attracts and the displeasurable repels, expectations of them become the determining causes of conduct that is chosen, or deliberate.

It must be realized, however, that in all impressions, external or internal, there is an inherent agreeableness or disagreeableness which determines their effect upon the psychic plane, and repeats itself in recollections or ideas of them. Accordingly pleasure and displeasure are very acute when the stimuli from which they proceed are very favourable or unfavourable to the life of the species or the individual, as for instance, the attraction of lust and the repulsion of physical pain. They accompany our sensations or feelings of psychic states, and are excited by the ideas that are formed of them. Thus we are pleased by courage and displeased by fear, whether as feelings of ourselves or as ideas of the nervous condition of another. Psychic revulsions, such as pride and shame, are acutely pleasurable or displeasurable. And by the differentiated or 'acquired' tastes which mark the complexity of civilization our pleasures and displeasures become multiplied in number indefinitely. The susceptibilities to which they appeal are mainly those of taste, sight, and hearing. But under psychic influence they take a further development. If a sensory impression implies superiority or inferiority it excites the *psychic* like or dislike of admiration or contempt. For superiority and inferiority are psychic stimuli since they connote the pride or shame of success or failure. Tastes which are accompanied by admiration or contempt are those which we discriminate as 'aesthetic'. Beauty is *excellence*, and its types vary with variations in standards of excellence.

So far of the psychic effects of *actual* stimuli. But it must be remembered that psychic energy may also be stimulated by recollections or ideas, that are unaccompanied by impressions of the senses. It may be an enthusiasm kindled by the brain as well as an emotion that spiritualizes a condition of the body. Thought plays upon spirit, as a harper upon his strings; consequently spirit, like a sensitive barometer, is incessantly fluctuating between rise and fall. The idea of a meeting with a beloved gives a flush of joy: the recollection of a parting a pang of sorrow. For we shall see that any nervous condition that has been accompanied by an idea can be recalled by the idea—or even by a trait of it—that is to say, can be *associatively* stimulated by it.

And, finally, psychic energy may be stimulated by cerebral nervous conditions. The brain is subject to neural states which resemble, with curious exactness, those of the physical nervous system. The presentation of a strange sensory impression or idea—one that cannot be connected with existing ideas—occasions a shock of surprise which is the precise counterpart of physical alarm. It is succeeded by the nervous state of doubt, corresponding to fear. This stimulates an effort of psychic resistance of which we are conscious as curiosity—an impulse to advance or explore, closely akin to courage. It may be followed by the effort of experiment—that is to say, by trial, guided by expectations of consequences—a phase of volition which may be distinguished as 'tentative'. When the brain is confused—not by a strange idea—but by the competition of two or more stimulating expectations, its neural state is that of hesitation, which may be compared, on the physical side, with dread. This is transformed into determination by an effort of psychic resistance, and we *choose*.

Accordingly, tentative and selective volition are guided by expectations—that is to say, by ideas of past consequences that are thrown into the future. In selective volition we *deliberate*; we balance competing expectations, and choose the most agreeable, or the least disagreeable. Our choice is, then, led by a fundamental instinct, to pursue the favourable and avoid the unfavourable. Amongst these expectations those of pride and shame may outbalance all others. Courage and curiosity are deliberate when they are inspired by a desire for the success that is followed by pride, or by an aversion for the failure that leads to shame.

It appears, therefore, that volition is the association of a physical effort with conditions of psychic resistance. The effort may be led by agreeable or disagreeable expectations. But it is 'free' in that resistance opposes itself to the instinctive response to a stimulus, and that the effort is primitively random. It may, indeed, remain so. In tentative volition we may make any experiment that suggests itself, however futile. In selective volition we may choose haphazard or by 'tossing up'. And, when willing simply opposes itself assertively to the sequences of instinctive or automatic life, it is quite unfettered, except in so far as its scope is limited by the range of the thoughts that occur to us.

When a mental state of surprise is suddenly dispelled there is an abrupt revulsion which is acutely pleasurable. This occurs if an idea that is strange from one point of view becomes familiar when regarded from another, when 'second thoughts' reverse its strangeness. The experience of this revulsion is our sense of the ludicrous, humorous, or comic which adds so immensely to the pleasure of life. Similar revulsions occur when hopes rapidly alternate with fears, as in betting and gambling, and the varied games which we term 'amusements'. At the present stage a reference to these pleasures must suffice.

Psychic energy is, then, swayed by the brain. But it exerts upon the brain an influence which is hardly less momentous. It forces the course of thought into emotional, as distinguished from critical, phases, in which the laws under which it acts are deranged in their action and lead to conclusions that are very different to those that would be formed 'in cold blood'.

We have seen that the psychic plane is characterized by muscular activities of its own—movements that are produced, not by stimuli that affect the physical plane, but by psychic energy in itself. These are the gesticulations and ejaculations, frowns and smiles, laughter, sobs, and tears which liberate emotion and *express* it. They relieve it by materializing it in action, which may possess no special character, as when little children 'jump for joy'. These movements are primitively automatic and unconscious; an infant's first cries are not *felt* by it. But ideas are formed of them, which bring them into harness, and we shall see that the gestures and utterances which express emotion come in the course of individual evolution to express ideas, and, together with efforts of trial, constitute the material from which conduct and speech develop. And we shall, moreover, trace to these psychic expressions, and the emotions that stimulate them, the remarkable mental and muscular activities that are termed 'imaginative'.

Let us summarize the leading features of this complicated discussion:

1. Our 'spirits' are an energy that is continuously generated by the spinal cord (or some part of it) in the form of motor excitement which takes one or other of two phases according as the stimuli from which it proceeds are favourable or unfavourable. We may characterize these phases as expansive and contractive. In consciousness they are happy or unhappy.

2. Its normal condition is expansive, and, when thrown into a contractive condition, it tends to revert to expansiveness.

3. This energy is independent of the brain. But through the action of the brain we become conscious of its activities as *feelings*, in which physical and psychic conditions are combined with ideas of them.

4. It possesses the properties which we term 'elastic'. Its phase is reinforced by a sensation or idea of accordant phase: it may be reversed in phase by one that is discordant: it may resist reversal; and it is subject to revulsion if a stimulus is changed or removed. Accordingly it acts as an elastic medium between the physical nervous system and the brain.

5. It converts neural states on the physical plane into psychic energies of which we are conscious as *emotions*.

6. By its revulsions it produces emotions of joy and sorrow: if revulsion follows the neural states which

are figured in idea as success and failure, it takes the forms of pride and shame.

7. It resists the incompatible, and hence may oppose itself, not only to nervous conditions of alarm and confusion, but to physical promptings generally, and even to the sequences of physical life. It transforms fear, dread, doubt, and hesitation into courage, anger, curiosity, and choice, and is the origin of emulation, industry, and morality. Resistance that stimulates a conscious effort is volition. The effort may be random, or guided by expectations of consequences.

8. Being of the same nature as the cerebral motor excitement which combines with sensory excitement to produce consciousness, it is affected by this excitement and affects it sympathetically, enhancing the agreeableness or disagreeableness of conscious impressions into pleasure and displeasure. These introduce new motives into life—the pursuit of the pleasurable and the avoidance of the displeasurable through their stimuli. Resistant effort, when guided by expectations of these consequences, becomes choice.

9. It activates special movements of its own—expressions of emotion from which has evolved much of our conduct that is expressive, much of our speech, and the imaginative capacity in thought and action.

NOTE A

It is permissible to theorize about the functions of the cerebellum, for they are admittedly one of the most discussed and disputed questions in physiology—a difficulty which, to use the words of Professor Bastian,[8] 'may still be considered to hold its place as a thoroughly unsettled problem'.

It has been demonstrated that lesion or removal of the cerebellum deprives an animal of its equilibrating powers. But this fact is not crucial, since these powers are also affected by the injury of the sensory (tactile) nerves of the spinal cord, of the corpora quadrigemina (optic tract) in the brain, and of the semicircular canals in the ear.

Rolando[9] compared the layers of the cerebellum to the couples of a galvanic battery and regarded it as a generator of motor force. Flourens (whose researches are the starting-point of the first clearly-defined notions regarding its use) considered that it possessed 'une propriété qui consiste à coordonner les mouvements, *voulus* par certaines parties du système nerveaux, *excités* par d'autres'. Weir-Mitchell believed that it was a source of energy to the other nerve-centres. Luciani[10] concludes that 'whatever the extent or degree of cerebellar lesion, the resulting symptoms are disturbances of voluntary movement'. It is recorded by Combette[11] that a girl in whom a cerebellum was completely lacking could stand and walk but 'se laissait souvent tomber'. All experimenters[12] agree in noting the profound lassitude which, in animals whose cerebellum has been excised, follows the execution of movements that are ordinarily quite unfatiguing.

Our theory is that this large ganglion is the principal seat of the energy which manifests itself in the emotions and will, and in the muscular and glandular activity that constitutes 'expressions of emotion'. This theory can draw support from eminent authorities. Professor Bastian notices that the homogeneity of

[8] *The Brain as an Organ of the Mind*, p. 49. See also on this point Ferrier's *Functions of the Brain*, p. 219; Foster's *Text-book of Physiology*, p. 1212; and Luciani's *Human Physiology*, iii. 480.

[9] See Morat's *Physiology of the Nervous System*, pp. 873, 874.

[10] *Human Physiology*, p. 481.

[11] Ferrier's *Functions of the Brain*, p. 180.

[12] Professor Sherrington in Schafer's *Text-book of Physiology*, ii. 910.

the structure of the grey nerve stuff of the cerebellum would accord with a theory that its function is merely to discharge energy [13] 'for the initiation of voluntary movements, in response either to volitional incitements coming to it from the cerebral hemispheres, or in response to unconscious impressions coming from the various sensory nuclei at the base of the brain and in the spinal cord', and he expresses an opinion that the cerebellum 'is a supreme organ (or motor centre) for the reinforcement and regulative distribution of outgoing currents'. In the judgement of Professor Sherrington [14] 'the main fact available from the study of the structure of the organ is its extraordinarily rich connexion with afferent nerves. The main phenomena displayed by cerebellar excitation and destruction are, on the other hand, not sensory but motor in kind'.

[13] *The Brain as an Organ of the Mind*, pp. 509, 520.
[14] Schafer's *Text-book of Physiology*, ii. 906, 909.

CHAPTER IV

THE BRAIN

A SENSATION, we have inferred, is the brain-stamp of an impression, illuminated by its own undischarged motor excitement. A durable replica of it remains as a record-idea, which becomes conscious as a recollection. We shall find that record-ideas can be disintegrated and recombined so as to form the complex ideas termed 'concepts'. It is, then, difficult to resist the conclusion that they, and the ideas that are derived from them, are material things—clusters of brain-cells, each cell representing a feature or *trait* of an impression. This may appear incredible, even grotesque. For ideas have an etherial quality. But they are only etherialized when they are illuminated by motor excitement. Record-ideas, whilst unsummoned, are not transfigured in this fashion, and may easily be conceived to be groups of brain-cells. It is, moreover, clear that ideas may be linked together in subconscious thought, and this is simply an automatic nervous process. And hypnotic experiments show that record-ideas may be acquired by subconscious [1] sensation, so that we may accumulate a stock of which we are unaware, which not improbably contributes to the strangeness of our dreams.

There is further evidence that record-ideas or recollections are material things which are systematically disposed in the cortex of the brain like books upon a library shelf. Post-mortem examination of the brains of those who have been afflicted with various forms of paralysis has established that inability to summon recollections

[1] Convincing reasons for this conclusion are given in Professor Moreton Prince's work on the *Unconscious*.

of a particular class is associated with the lesion of a particular cortical area of the brain; that is to say, record-ideas have actually been localized. Injuries to the cortex of the occipital lobe render the sufferer unable to recognize objects by sight: they destroy the records that are needed to complete his sensations of sight; and the class of objects which he cannot recognize varies with the situation of the injury. In like fashion the cortex of the temporal lobe is shown to be the depository for records of hearing, and that of the parietal lobe for records of muscular movement. An injury to a patch in this lobe, of the size of a hazel-nut (called 'Broca's convolution'), is followed by inability to utter words[1] voluntarily. We cannot, then, avoid the conclusion that a record-idea is a cluster of specially modified or impregnated brain-cells; Professor Moreton Prince suggests for it the expressive term 'neurogram'.

How, it may be asked, can currents of sensory excitement imprint themselves permanently upon brain-cells? We have no such experience as would enable us to answer this question: we are in the case of children wondering over the intricacies of a watch. In photography, vibrations of the aether record themselves by effecting chemical changes—although in a form which gives no clue to their real character. This affords us a distant and vague analogy which may enable us to comprehend, in a fashion, how impressions of the external senses, such as those of surface, colour, sound, and smell, may make changes of nervous substance which would stereotype their effect. But we are thrown into confusion if we attempt to think how permanent imprints can be formed of the movements of our limbs, of neural states such as those of hunger, fear, and success, of psychic conditions of expansion and contraction, of efforts and of their intensities, of the relations of one thing to another, as that money is *in* the pocket, or

[1] A summary of these correspondences is given in Note B at the end of the chapter.

PLATE I

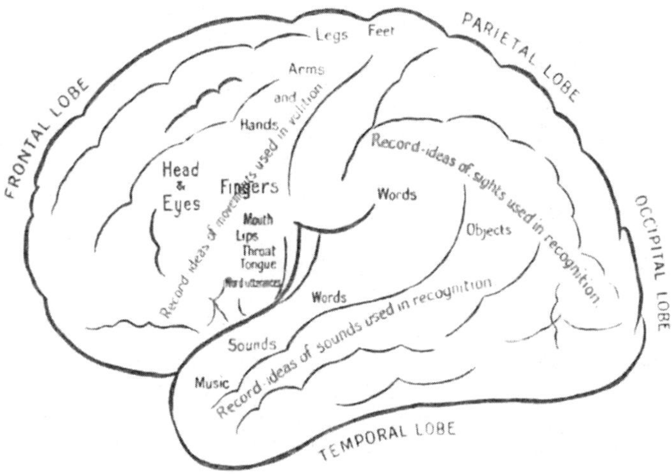

Diagram of the left hemisphere of the brain showing, generally, the areas appropriated to record-ideas of different classes, as indicated, so far, by various forms of paralysis. [**Modified from a drawing in** ***The Brain and Personality*** **by Professor W. E. Thomson.**]

sugar mixed *with* tea. We can certainly recollect all these experiences. This warrants us in assuming that they are imprinted upon the brain, although we cannot conceive the precise nature of the record. But we shall find reasons for inferring that it is an elaborate complex of numerous traits, and that it presents itself, in many cases, not as a fixed image, but as a rapid series of changes, so that it may be likened to a musical phrase, or cadence, rather than to a chord.

The brain consists of myriads of nerve-cells connected by an intricate network of fibrils, which provides them with a means of communicating with one another, or of linking themselves into groups, in however complicated a fashion. Each hemisphere includes two layers of grey ganglionic tissue, one concentrated in various masses[3] at its base, the other spread over its surface, and forming the cortex. Between them lies a mass of white nerve-tissue which constitutes the most substantial portion of the hemisphere. It is mainly composed of fibrils that radiate from the basal to the cortical ganglia, and bring the two layers into close connexion; but it also includes transverse fibrils that establish communication between the various areas of the cortex. It is prolonged into branches which connect the basal masses of ganglia, and the hemispheres, with one another.[4] We have, then, two aggregates of ganglionic nerve-tissue—one basal, the other cortical—which are distinct, but are united in the closest intimacy.

Now this twofold distribution of the brain-cells has a remarkable coincidence with the twofold character of our ideas as *sensations* and *recollections*. These are similar and yet different. The former are vivid, the latter of a fainter and more general character. We have distinguished them respectively as *impression-*

[3] The principal of which are the *optic thalami*, the *corpora striata*, the *corpora quadrigemina*, and the *corpora geniculata*.

[4] And vast numbers of connecting fibrils run to the cerebellum and spinal cord.

ideas and *record-ideas*—the former being ideas that are in course of formation through the process of sensation, and the latter, ideas that have been formed and put away. These, we have seen, are assembled in the cortical layer. Regarding the locality where impression-ideas are formed we have no such definite knowledge. But there are facts which appear to indicate that it is in the basal masses of nerve-cells. For observations [5] made of fishes, frogs, and pigeons whose brains have been excised show that, so long as the basal ganglia remain, they can see, after a fashion, and can feed themselves, whereas if these ganglia are removed, along with the superior hemispheres, they lose all power of conscious action and become mere automata under the unconscious (reflex) action of the spinal cord. Indeed a fish that has been deprived of its upper brain but left with the basal portion, behaves in much the same fashion as a normal animal. And this conclusion is borne out by anatomical considerations which shows that the two largest masses of basal cells—the optic thalamus [6] and the corpus striatum—are intimately associated, respectively, with the reception of outside impressions and of impressions of internal or bodily activity. We have accordingly some grounds for an inference that it is in the basal cells that *sensation* takes place, in the course of which impression-ideas are formed out of the sensory excitement that is aroused by stimuli, and that records of these ideas are made and stored in the cortex of the brain.

[5] Details of the experiments are given in Ferrier's *Functions of the Brain* (second edition, p. 109), and in Schafer's *Text-book of Physiology*, ii. 699–702.

[6] There is great uncertainty regarding the functions of these organs. But of the former Sir E. A. Schafer writes: 'The sensory tracts are ultimately traceable to the *thalamus*, which must be regarded as a great sensory centre' (*Text-book of Physiology*, ii. 768). And Professor Bastian concludes that 'the *corpora striata* are great motor ganglia, concerned with the execution of voluntary movements. The destruction of one of them produces a complete loss of voluntary power over the limbs on the opposite side of the body' (*The Brain as an Organ of the Mind*, p. 566).

PLATE II

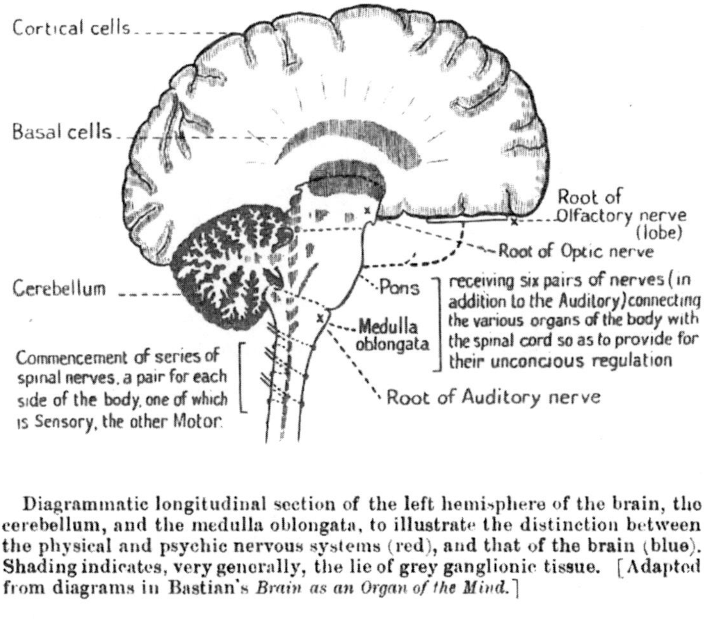

Diagrammatic longitudinal section of the left hemisphere of the brain, the cerebellum, and the medulla oblongata, to illustrate the distinction between the physical and psychic nervous systems (red), and that of the brain (blue). Shading indicates, very generally, the lie of grey ganglionic tissue. [Adapted from diagrams in Bastian's *Brain as an Organ of the Mind.*]

The operations of the brain involve the establishment of connexions between ideas. For it is clear (1) that a sensation is not completed until the impression-idea has connected itself with amplifying records, (2) that complex ideas are formed by the connexion of certain traits of one record with those of another, and (3) that in thought ideas are connected into a running series. These connectings must have a cause in some attractive force. Self-analysis shows very easily that one idea summons another in thought if the two have been previously associated together, or if there is a resemblance between them, and these facts have been generalized as the principles of 'contiguity' and 'similarity'. But these principles merely represent *consequences*: they contain no explanation of the cause which underlies them. This must obviously be searched for in the very elements of life. Now one of the most fundamental of living forces is the attractiveness of the familiar and the repulsiveness of the strange. Experiment shows that no living creatures are too simply organized to be unaffected by them, and we may reasonably conclude that familiarity is the ultimate base of gregarious life. The familiar and the strange are the origin of peacefulness and fear, two of the most influential of our feelings. They affect our social life in drawing us towards acquaintances, and setting us against strangers. Familiarity is the attraction which binds us to our habits and our tastes, and even renders our food [7] easily digestible.

It is manifest that ideas which have been associated together in sensation or thought have become familiarized. And it is hardly less obvious that two ideas which resemble one another are familiar. For similarity involves the possession of a trait, or sensory element, in common; things are *like* one another because, in some respect they are the same. If the trait in common

[7] It seems to affect the action of some inanimate forces: the electric current is apparently susceptible to it.

is one of *relationship*, we speak of similarity as 'analogy', and this, as we well know, connects one idea with another. We may, then, distinguish familiarity as resulting either from previous association, or from identity or sameness of trait. By an effort of thought we may make a generalization which will assimilate association with identity. For things that have been familiarized by past association possess a sameness of rhythm or time, since they have fallen either together or in sequence. And if, as is believed, external stimuli are vibratory, we can go further and assimilate trait with rhythm, for identity of trait would be identity in vibratory rhythm. Casting our reflections still further back, we may seem to discern a Law of Harmony under which vibrations attract one another when they are identical in internal or external rhythm. But we are abandoning the thread of inference for the gropings of speculation, and for practical purposes we may rest content with the conclusion that familiarity is of two kinds—familiarity of rhythm and familiarity of trait. We must, however, bear in mind that by familiarity of rhythm two ideas may be bound together either coincidently or in sequence, according as they have been associated in sensation, or in thought, simultaneously or successively. In the former case one idea embraces the other, so that, if one is recalled, the other accompanies it so closely that the two are merged together. In the latter case one idea is linked to its successor, so that it is *followed* by it, and the two are distinguishable from one another. A memorial chain consists of a series of pairs linked together in this fashion. And we shall see that, when recalled, the pairs may present themselves in reversed order.

Now the attractive forces of familiarity of rhythm and of trait are plainly identical with memory and intelligence. A plant recalls its botanical name because the two have occurred successively in reflective experience: this is memory. It may summon a thought of the Order to which it belongs because it possesses traits that are identical with those of the Order: this is intelligence, which may be correctly and convincingly defined as the appreciation of samenesses that underlie differences. It follows that memory and intelligence, which we figure as causes,[8] are in reality consequential conditions of the brain.

We must not regard the effect of familiarity as moving brain-cells into physical contact. It would produce its results if it switched them into connexion through nerve-fibrils, myriads of which are interlaced with the nerve-cells of the brain. When familiarity of rhythm is the connecting force, the two ideas remain distinct, and we are aware of the establishment of a connexion between them. But when the connexion is made by familiarity of trait the process eludes us. For the traits that are identical are *unified*, and lose distinctiveness, just as two notes that are in unison blend into one. If the idea of a walk calls up that of a hat in association, we are aware of the memorial connexion. It may call up the idea of a bicycle because walking and bicycling are identical in being means of progression: in this case we do not realize that two traits have been connected. Their connexion eludes us because they have merged into one.

Let us now see whether these laws will enable us to analyse the process of sensation, and the formation of record- and concept-ideas. An impression of sight, on reaching the brain, becomes an impression-idea which is illumined by the motor excitement that the impression has generated. But it would be a meaningless representation of flat outlines and colours were it not connected with recollections which complete the image by adding details that are not perceptible to the senses. It would be so connected, were it unified with a record-idea, or recollection, of itself or its traits. This, by

[8] Just as the chemistry of a century ago attributed the condition of heat to the action of a mysterious substance, called Phlogiston.

familiarity of rhythm, would draw in train other recollections with which it had previously been connected. Unification would occur through a fibril connecting nerve-cells in the basal ganglia that form the impression-idea, with nerve-cells in the cortex. Association would be effected through fibrils passing across the cortex. Accordingly sensation involves the use of intelligence as well as memory. It is by intelligence that an impression-idea is connected with a record-idea which is its generalized counterpart—that is to say, that it is 'recognized'. But since this connexion is a unification, it escapes notice.

Perception involves differentiation and dissociation as well as unification and association—that is to say, it involves the separation of elements that are unfamiliar in trait or in rhythm. By the senses of touch and sight we receive a number of impressions simultaneously. We may feel many things, and see many shapes and colours at the same time. They are isolated from one another because their stimuli are of different characters. If they are not so differentiated our sensation is of unbroken continuity, as of a smooth surface, or an unclouded sky. If differences occur, not simultaneously but successively in a sequence, they are *changes*. An incident or happening is a succession of changes. If these changes are of internal touch in our limbs, they are movements. These, if progressive, are accompanied by changes in the visual size and perspective of the objects around us, and recollections of the two sets of impressions are associated. Consequently if we see changes in the size or appearance of objects we perceive that they are moving, unless we are moving ourselves. If the size of a motor-car rapidly increases, we perceive that it is moving towards us. We are, however, anticipating the subject of *perception*, which will be separately discussed in a later chapter.

If we close our eyes after looking at an object it may remain for a short period 'visualized'—that is to say, vividly presented to us. More or less rapidly, however, it fades away and its place is taken by a recollection—a thing which is less distinct and is, we feel, of a different quality. But it endures, whereas the impression-idea fades away. It appears, then, that the basal cells are only temporarily impressed in sensation, so that, to use an illustration from photography, they remain ready for fresh exposures. On the other hand, recollections or record-ideas are more or less permanent replicas, or 'photographs' of impression-ideas, formed by the transmission to cortical cells [9] of the characters that are impressed upon basal cells in the process of sensation, just as a photograph may be taken of a photograph. They register the nervous effect of a stimulus as the pittings of a phonograph record register the waves of sound that have been thrown upon it. The effects of sensory excitement are not merely impressed, but are stereotyped upon them. The basal cells are, as we have seen, very intimately connected by fibrils with the cortical cells, and, although we may appear to be introducing an extraordinary complication, it is quite consistent with such knowledge as is possessed of the transmission of nervous energy to suppose that a basal cell can transmit to a cortical cell the conditions into which it has been thrown by sensory excitement. It is beyond doubt that nerve-cells can communicate the energy of excitement to one another—indeed, it is in this fashion that impulses are transmitted to various portions of the nervous system.

The character which is thus stereotyped upon a cortical cell would determine whether, when affected by motor excitement in the course of thought, it would vibrate, so to speak, expansively or contractively, and

[9] It is a significant fact that most of the cortical cells are anatomically and chemically peculiar in containing few or none of the 'Nissl granules' which generally characterize the nerve-cells of the spinal system. These are known to be used up during nervous activity and to be regenerated during rest. It may be inferred that the activity of the cortical cells is not exhausting and transient, but durable.

the intensity of its vibrations of either phase. But we must not overlook the effect of sympathetic action and reaction between it and psychic excitement. This may modify the intensity and even the phase of its vibration.

The impressions which the brain receives in sensation are merely those of features or traits, such as textures, colours, and outlines. These are combined into individuals or objects and invested with psychic attributes by processes of individualization, differentiation, and association which will be discussed in the chapters following. Accordingly perception involves thought as well as sensation, which of itself tells us very little indeed. In chapter ii we have already reviewed the general character of our sensory impressions. We may add here a few comments upon them.

We are sensible of our psychic states in our feelings.[10] These are impressions of motor excitement, to which physical elements may be added. Consciousness involves motor excitement, and we may accordingly be impressed by it and form an idea of it—as a feeling. But motor excitement possesses no traits except those of phase, as expansive or contractive, and intensity, as mild or severe, and we can, therefore, form no distinct ideas of feelings in themselves. Consequently they cannot be recollected apart from the effects of their stimuli or the actions which accompanied them. We cannot remember a particular fit of anger apart from the incidents which aroused it, or the actions to which it led. Actions involve impressions of touch which particularize the feelings that motived them. So again with psychic pleasure and displeasure. No recollections of them can be summoned apart from their causes, or the movements by which they expressed themselves. Nor can we think of consciousness apart from the ideas that are conscious.

As regards our environment, sight appears to be the

[10] In French termed 'sentiments' and thus at once identified and contrasted with sensations.

most informing of our senses. But that of touch really tells us infinitely more. Combined with feeling it affords us ideas of resistances which we conceive to be varying conditions of *substance*, liquid when we can, solid[11] when we cannot break through its surface; ideas of texture, as of rough or smooth, of shape, contour, and dimensions. It impresses us with *number*, since we are sensible of the separateness of several objects that are touched simultaneously. It tells us of relative as well as of absolute temperature, for, since our ideas of this are mostly derived from pressure exerted upon the nerves by the expansions and contractions of body-tissue, tepid water seems hot or cold according as we have just withdrawn the hand from cold or hot water. For it causes expansion in the first, contraction in the second case. And it is apparently by touch that we feel internal conditions, such as hunger, and the physical changes which accompany the emotions—to be noticed more particularly in chapter xi.

The fact that touch is the most archaic of the senses has a momentous consequence. We instinctively regard the tangibility of an object as its substance or *essence*, whereas impressions of smell, colour, and sound that arise from it are its *qualities*. And, where possible, we give a tactile existence to impressions that reach us through the other senses. Many of our impressions cannot be materialized in this fashion—such as light and darkness, heat and cold, and our internal conditions. We are not content until we can consign them to *causes* or antecedent impressions which are tangible or can be figured as tangible. We search indefatigably for the causes of our ailments. We materialize the changes that occur in our psychic feelings by connecting them with such causes as our food and drink, the possession or lack of money, or the effect upon us of the actions and words of others. These are tangible: they involve internal touches.

Sight gives us ideas of colours and outlines, also of

[11] Our primitive idea of the gaseous condition is derived from smell.

extension, to which we add depth by the association of the size and perspective of objects with their *distance*—that is to say, with the amount of effort that would be required to reach them. The field presented to our eyes is constantly changing: objects grow larger or smaller as they approach or recede; if they are in transverse movement they displace other objects as they cross the field. Hence we obtain clearer ideas of motion than touch alone can afford us.

Through hearing we obtain recollections of external rhythm, of notes (pitch), of timbre, and of the melodic sequences in which successive notes are distinguished by differences. We can recollect chords—that is to say, we distinguish several notes that are sounded simultaneously, though not, as a general rule, very clearly.

The delicate nerve endings of the ears and eyes have enormously increased the range [12] of our susceptibilities. But touch remains the most fruitful source of ideas of our environment. The writings of Helen Keller [13] show how rich in ideas may be a life to which impressions of touch alone contribute. Blind and deaf from the age of eighteen months, her only conception of sights and sounds were derived from impressions of finger-touches—of raised letters or a companion's fingers 'spelling into her hand'. Yet she writes with vivid appreciation of natural beauties—of sunsets, of flowers, of the songs of birds, although her acquaintance with them was purely symbolic.

This forces us to reflect upon the reality of our ideas. How far can we trust to them for information as to the real conditions of things outside us and of our own bodies? Ideas, as we have seen, are the products of sensory excitement, and can resemble the causes of this excitement only if sensory excitement resembles its

[12] It remains, however, strictly limited. We cannot detect vibrations of sound which are less than 16 or more than 38,000 to the second. We cannot translate into light the ultra-red vibrations of heat, or the ultra-violet (actinic) vibrations.

[13] *The Story of my Life*, *The World I live in*, and others.

stimulus. This appears to be in the last degree improbable, and we must then, it seems, think of our ideas as symbols, which resemble reality no more than a catalogue of pictures resembles the pictures. Common sense revolts from this conclusion. It appears impossible that our impressions of the world should give us no information as to the constitution of the world. But we may learn from Helen Keller's experiences how greatly we may be deceived. To her a finger touch seemed to be really a colour or a sound. We cannot, then, claim that our ideas must be more than symbolic because they *seem* real.

Do we, then, know nothing that is real in regard to our environment ? Is science a mere playing with symbolic phantasies ? Assuredly not. For our perceptions, however unlike reality, give us correct information in regard to reality, namely, that it is a series of changes and persistencies, which possess identities and can therefore be classed, and are commonly linked together in sequences. We regard the sequences as of cause and consequence ; we express the regularity of their succession as a 'law', and use these laws in the process of inference. A sequence may be rhythmical. In this case we can make it the subject of calculation—for arithmetic, as we shall see, is the science of rhythm—drawing such inferences as those which assure us of the movements of the heavenly bodies and of the vibratory nature of sound, light, and heat. And if changes and persistencies can be measured, we obtain relationships which enable us to reason quantitively concerning them, although their precise character lies beyond the scope of our perception.

If, however, ideas differ in nature from the stimuli which they symbolize, how is it that they can touch the physical nerve-system as if they were actual stimuli ? That thought may affect us physically is clear beyond doubt : the idea of a danger may set us trembling with fear ; the thought of a meal may make the mouth water, exactly as may its sight or smell: even dream-fancies produce physical effects. How can these symbols affect the nerves as if they were actualities? Through rhythmic association. They have been associated with actual stimuli in rhythmic coincidence and are therefore linked, in rhythmic sequence, with the movements that are the muscular consequences of the stimuli. A sequence that succeeds two elements in coincidence can be initiated by either of them. An injury is accompanied by the idea of it, and induces nervous consequences which include a movement of shrinking: hence the idea [14] or thought of it induces these consequences. How then, it may be asked again, can the physical plane be affected by the idea of an impression which has never been actually experienced—as, for instance, by an incident of a ghost story? Because the idea is unified with one of actual experience by a sameness of trait. The mysterious is strange and dangerous, and excites the consequences which they have excited. An idea which was altogether unconnected with experience would leave us unmoved. So novel a thought can hardly suggest itself to the most exuberant imagination. But the general truth of this conclusion is recognized in the line—'He jests at scars who never felt a wound.'

This process of 'associative stimulation' enables the brain to control the psychic as well as the physical nervous system, so that an emotion can be recalled without repeating the nervous processes that were involved in its original evolution. An idea of a mother arouses the feelings of love and respect, because these emotions have accompanied actual experiences of her kindness and authority, that have been associated with ideas of her. So ideas of antagonism, success, or excellence, having been rhythmically associated with psychic

[14] The effect of association is illustrated by a well-known psychometrical experiment. One who has been subjected to a succession of electric shocks, each accompanied by a note of music, will wince when the note is sounded, although untouched by the current.

resistance and the expansion of triumph, stimulate these conditions by recalling them. We have seen that ideas become *passively* pleasurable or displeasurable through the sympathetic effect of psychic excitement. But they are *actively* stimulating because they are memorially (rhythmically) associated with nervous conditions. In experience they have become paired with nerve-cells in functional sequence, and they stimulate them neuro-serially by recalling them to activity. Their stimulating effect may be likened to a *pull* as contrasted with a *push*.

The salient points in this chapter are:

1. That ideas, whether impression, record, or concept, are material things, and may be figured as clusters of brain-cells, each representing a trait, or feature of an impression.

2. These cells and clusters are drawn into connexion with one another by the familiarity which results from coincidence or sequence in rhythm and from sameness of trait, the action of the former being memory, and of the latter intelligence.

3. Perception involves the union of the traits of an impression-idea with record-ideas possessing identical traits which draw in train with them other record-ideas with which they are connected by familiarity of rhythm.

4. We perceive changes through differences.

5. Ideas are merely symbolic, and do not represent the actual characters of things, but they give us reliable information as to similarities and differences, coincidences, and sequences, which exist in the world around us.

6. The brain stimulates the physical and psychic nervous systems through associations that have been established in experience between nervous conditions, or muscular movements, and ideas of these conditions and movements.

NOTE B

In one phase of paralysis the sufferer suddenly loses the power of recognizing objects. He may see houses or faces,[15] but they mean nothing to him because he cannot associate them—that is to say, the visual impressions are not completed by recollections. This failing is attended by a lesion in a particular cortical area of the *occipital* lobe of the brain. Still more remarkable are the effects of the various phases of language paralysis. The sufferer, while continuing to *see* written or printed words, may find himself unable to *read* them.[16] They signify no more to him than if they were printed in Russian characters, or backwards. He may be able to recognize letters but not words, or words but not single letters, or may retain his command of figures whilst losing that of words and letters. More extraordinary still, he may retain command of one language, but lose that of another: in one case a patient could not read his mother tongue with understanding, but could still comprehend Latin and Greek. These failings are associated with lesions in other particular cortical areas of the occipital lobe.

So also when hearing is paralysed, ability to hear sounds may continue. But they cannot be recognized because records are missing wherewith to identify them. In different phases of hearing-paralysis one may lose the power of recognizing ordinary sounds, the sounds of words, or of music, and in each of these cases there is a lesion of a definite cortical area in the *temporal* lobe of the brain, which is evidently the locality in which record-ideas of sound are, so to speak, 'filed' or put away.

In another phase of paralysis the sufferer loses the power of *voluntarily* moving his lips, or one or other of his limbs, although he may still be able to move them involuntarily under the stress of a powerful stimulus (such as sudden fear) which affects the spinal cord. That is to say, he has lost the power of using

[15] In some cases those who have lost powers of recognizing, say, the faces of their acquaintances, or of printed words, can still recall them in idea, if asked to do so. Their injuries are of a different order, and have destroyed, not their record-ideas, but the channels by which impression-ideas are connected with them.

[16] Interesting details of these cases are given in Dr. Hinshelwood's *Letter, Word, and Mind Blindness*, and *The Brain and Personality* by Professor W. H. Thompson.

the *ideas* of movement which invariably precede an effort of will. The command of voluntary movement that is lost may be that of uttering words, when there is the paralysis known as 'aphemia'. Each of these failings is accompanied by a lesion in a particular cortical area of the *parietal* lobe.[17] Record-ideas of movements are evidently arranged here in an order which is the reverse of that of bodily structure—those of leg-movements being at the upper, and those of movements of the tongue and lips at the lower, end of a band that runs diagonally across the lobe. The actuality of these records is demonstrated very curiously by the fact that ideas of moving a limb may persist after the limb has been amputated.

It may be added that record-ideas of the sight, sound, or utterance of *words* are generally all deposited in the left hemisphere and are, then, altogether abolished by lesions in this hemisphere. But those of other sights, sounds, and movements are located in both hemispheres, those of each hemisphere affecting the use of the opposite side of each eye, or a limb on the opposite side of the body, so that, in respect to them, lesions in one hemisphere only produce partial paralysis (hemiplegia).

[17] It has been shown by experiments with animals (chiefly monkeys) that injuries to the parietal lobe check the muscular responses which are ordinarily excited by the electrical stimulation of a sensory nerve.

CHAPTER V

THE EVOLUTION OF CONCEPTS

We can explain the formation of the complex ideas termed 'concepts' if we assume that impression- and record-ideas may be disintegrated, so that the elements of which different impressions and records are composed may be drawn into association, or be unified, when they are familiar with one another in rhythm or in trait. At first sight it seems incredible that ideas should be built up by the action of certain nervous susceptibilities, although we appreciate the consequences of these susceptibilities, and speak of them in such terms as *association* and *dissociation*, *unification* and *differentiation*. But their action is hidden from us when it affects, not whole ideas, but the elements of which ideas are composed. We are not conscious of the disintegration of record-ideas, and the selective reconstitution of their fragments, and can only judge of their actuality by inference. There are mental processes of which we are unaware—dark chambers in the brain's workshop that are not illumined by the light of consciousness.

We are conscious of a memorial process when it links two ideas in sequence, as when the name of a plant is associated with its appearance, or a number of ideas or words are strung together in a chain. In these cases familiarity of rhythm consciously manifests its activity. But we shall see, as we proceed, that it also affects us subconsciously, and that, below the surface which is lit up by consciousness, it is constantly weaving physical, psychic, and mental processes—including our movements—into a web which preserves established connexions. And we can infer that it affects the elements, or traits, of which record-ideas are composed, as well as the record-ideas themselves.

So again with the familiarity of trait which manifests itself as intelligence. Its action is mostly subconscious since it generally uses the elements of ideas, not whole ideas, as its materials, and when two elements are unified their distinctiveness is, of course, obliterated. The samenesses, which lead to unification may, however, present themselves consciously—as when we classify a thing by unifying it with a group—so that their apprehension has been a process of partly conscious, partly subconscious evolution which has been co-extensive with man's progress from primaeval savagery. The unifications or classifications that have resulted are, however, handed down to succeeding generations as memorial associations, or coincidences, between words and the attributes of some particular illustrative objects, so that those who learn them do not realize the process by which they were evolved. In thinking of 'elasticity', for instance, we do not figure to ourselves that this conception arose from the unification of a peculiar property that is possessed by several substances; we regard it, primarily, as an attribute of the substance used to illustrate it—indiarubber. So memory, in transmitting the fruits of intelligence, obscures their origin, and it requires a considerable effort to shake off its obsession, and to reconstitute[1] concepts as they were originally formed.

We must think of a record-idea as a cluster of brain-cells, each of which represents a trait of the impression from which it was formed. The cells are held together by familiarity of rhythmic coincidence, for they have been impressed simultaneously—just as the details of a picture at which we have looked are held together in memory, so that we recall them together. By subse-

[1] One may frequently be assisted in this endeavour by the archaic significance of words, which expressed the real character of ideas with infinitely greater precision than remains in the conventional meanings that have subsequently become attached to them.

quent perceptions of the same, or of other objects, these traits are differentiated from one another as being either *essential* or *accidental.* The former are universal or invariable—that is to say, they occur in *all* impressions which in perception are unified with the cluster as being of the same object or of the same kind of object —or, stated differently, they *always* occur in these impressions; they are, therefore, bound to the cluster by continuous rhythmic coincidence. They are also *peculiar* to the cluster, in that they occur in no impressions other than those which are unified with it in perception: they therefore differentiate it from all other clusters. Consequently they cannot be isolated or detached from it. So it comes that we form no separate ideas of traits that are both invariable and peculiar. The scent of lavender, for instance, is invariable and peculiar: it cannot be detached from the plant. If pink was invariably and peculiarly the colour of roses, we should have no name for it but 'rose colour'. If a brown colour was possessed by all dogs, and only by dogs, we should speak of it as 'dog colour', as we do of 'terra cotta'.

When a trait is either variable (that is to say, occasional) or common to impression- and record-ideas of different kinds, it is *accidental.* It is occasional if it sometimes occurs and at other times does not occur in impressions which are unified with the record-idea. It is therefore discontinuous in coincidence with the cluster, and can be detached from it. So redness can be detached from recollections of roses because there are roses which are white or yellow. But even when traits are invariable in occurrence, they can still be detached if they are shared with other different recollections. For in this case it is discontinuous with the cluster in rhythmic coincidence: it occurs apart from the cluster. Thus all dogs are hairy, but since hairiness is also a trait of other animals, it is an accidental trait, and can be detached from the recollection of a dog.

In this fashion essential traits are differentiated or isolated from accidental traits in the same cluster by strangeness of rhythm, while each accidental trait is isolated from all other accidental traits by its inherent strangeness. All the traits so isolated which occur in different record-ideas are then *generalized* by being unified; since each group of essential traits and each accidental trait is drawn to its affinities by familiarity of trait. So generalized they become trait-classes. When we speak of a *rose* we refer to a general group of essential traits, and when of the *redness* of the rose, to a generalized accidental trait.

A process which runs alongside of that just described is that of individualization, by which we form concepts of ourselves, other persons and things, as standing apart from their environment and possessing an independent existence. The most fundamental of all individualities is our own. We are differentiated from our surroundings by discontinuity of rhythm, for we can move independently of them. And we possess a persistent unity—that is to say, a continuity both in traits and in rhythm. The continuity is threefold, of substance, energy, and thought, and accordingly we think of ourselves as consisting of body, spirit, and mind. The three are linked into close association because they are coincident in rhythm. If we add an idea of our outward appearance, which masks or reveals our feelings and thoughts, individuality becomes *personality*.

We individualize, as other persons or things, essential traits, or groups of essential traits, which are discontinuous with their environment. Most 'individuals' are also distinguished by the possession of either substance or energy or continuity of existence. Substance, as we have seen, is tangibility; movements of the body involve sensations of touch, and we therefore individualize them, as in *blow* and *kiss*. Parts of the body are individualized, since they can be divided from the body. As manifestations of energy we individualize the various emotions and the will. Being continuities, trait-classes are individualized as abstracts—*redness* for example—since they continue apart from the individuals to which they belong. But things which possess neither substance, nor energy, nor continuity are individualized if they are discontinuous with their environment—a rainbow, or a note of music for instance.

If objects so individualized manifest the movements which we associate with life and motive, they are endowed by association with life and motive. These are qualities which we cannot perceive. If they are inanimate, they may still be endowed with purpose should they contribute to the fulfilment of a motive. Such ideas as those of a *chair* or a *book* have no meaning apart from their implication of purpose. Hence man is disposed to see motive or purpose in all around him, and is quite ready to believe that animals, and even objects, may affect him mysteriously.

Our concept of an individual is termed a 'particular' idea, marked by a proper name, or by the definite article. But it arises in fact from the coalescence of the features which enter into a number of particular ideas. If we see a dog in three positions we form three different impressions and recollections of it. But certain features are invariable and peculiar, and occur in all impressions of the animal. They are isolated by the discontinuity of their accompaniments, are unified by familiarity of trait, and become a generalized concept with which we unify subsequent impressions of it, and accordingly 'recognize' it. When some of these invariable and peculiar traits occur in several distinct individuals, they are similarly unified into a still more generalized[2] concept of *kind*. This represents continuity of certain essential traits but discontinuity in individuality, for the group of traits remains constant, but there are variations in the individuals that are unified

[2] Plato's 'general ideas' have then a physical existence in the brain, although not in the world outside it.

with it. We recognize,[3] or unify, the nature of an animal or object by connecting the visual or tactile impression of it with its *kind* group.

The concept of a particular individual may be formed subsequently to that of the kind to which it belongs. In this case it is isolated from its kind by the possession of distinctive or peculiar traits. And it is to be observed that ideas of kind may be enlarged from species into genera, or wider classes, by being unified with other ideas of kind which possess identical traits. Thus the kinds *dog* and *horse* are amalgamated in the kind *quadruped*; the kinds *courage* and *sincerity* into the kind *virtue*.

There is a further complication. The essential traits which enter into the idea of a kind may be dissociated from them if the kind be regarded as a substance or an energy. For in this case all its other traits become accidental. So from the traits of the kind *stone* can be evolved the essential trait-class *stony*, and from this the essential abstract *stoniness*; from the kind *joy* the essential trait-class *joyous*, and the essential abstract *joyousness*. Trait-classes and trait-abstracts can accordingly be formed by the disintegration of ideas of kinds, as well as directly, by the isolation of accidental traits.

It may seem that this description of the processes of disintegration, isolation, generalization, and individualization, which transform crude record-ideas into concepts, is entitled to no more credit than would be given to the wildest of speculations, since we are actually unconscious of these elaborate operations of the brain. But reason enables us to ascertain what we cannot perceive by the senses. We can infer the existence of causes by the consideration of consequences. If, for instance, we find that a collection of flowers has been arranged by their

[3] The process of recognition is immensely facilitated by the use of words. If a child is told that an animal is a dog, he classifies by the name, without thought of its identities. For the sound 'dog' is associated with the kind-idea as a distinctive feature.

families, genera, and species, we conclude that there has been a process of classification. Accordingly if by analysing our concepts we discover that they differ from crude impressions in possessing certain elaborate characters, we may infer that their elaboration has followed lines which will develop these characters.

For the sake of simplicity the illustrations which we have been using have been drawn, for the most part, from impressions of the external senses. But it must be realized that these contribute much less to the formation of concepts than do our impressions of internal touch, of neural conditions, of psychic activities, and of consciousness. We shall see, for instance, that feeling enters into the evolution of ideas of solidity and weight, and that some of our most refined conceptions are drawn from our experiences of psychic changes.

And in the evolution of ideas of traits there is a further development of great importance. The traits which we have been considering are those of *character*: that is to say, they can be possessed by an individual independently of other individuals or the environment. But a vast number of traits are of *relationship*, that is to say, are derived from the association of two or more things in certain fashions. The idea of a son implies one of a father; a cause cannot exist without a consequence; there cannot be an accusative without an action; a thing cannot be 'more' or 'less' except by comparison, or exist in place or time without ideas of place and time. We have words to express some of these relationships—'sonship' and 'causality', for instance. But most of them are unnamed—those signified by prepositions, for example, although we can *think* of 'inness' in the abstract. Sameness in relationship is commonly termed 'analogy': there is an analogy between speaking and eating, because they both *involve movements of the lips.* Relationships may be numerical or quantitative: two numbers are connected in relationship by being equal, or in proportion, by a contrariety which is expressed, respectively, by *plus* and *minus*, or by their contributing to form a total; and it is by the use of such relationships that the mathematician is able to ascertain actual numbers or quantities from *data* which at first sight appear to give no indication of them.

Relationships are the links which bind the various ideas or words of a thought or sentence together, and it is through them that thought evolves from the simple into the complex. Thus our illustration of analogy in relationship may be elaborated into 'movements of the lips *and* tongue, *in* accord, *under* stimulation, evading *consciousness*'. Grammar is the art of expressing traits of relationship by the amalgamation or modification of words or their orderly arrangement, so as to formulate a thought. This involves the connecting of an individual or abstract—the 'subject'—with a condition, active or passive, that is 'ascribed' to it. The condition, as we shall see, is that of *consciousness*—consciousness of being the same as, or of being associated with, an individual, of being unified with a kind or trait-class—that is to say, of being *classified*—or of being unified with an active or passive feeling. Both subject and condition are defined and expanded by traits that are used as attributes or complements. The processes of perception and thought will be discussed in chapter vii, but it is necessary to anticipate in some degree if we are to render the nature of some concepts intelligible. The condition is connected with the subject to which it is ascribed (or of which it is 'predicated') by the verb—a concept which it is particularly difficult to analyse. Under careful scrutiny, however, it resolves itself into three elements—personality, consciousness, and a feeling—bound to one another by unification, since consciousness is, in fact, identical with personality, and, in apprehension, a feeling is identical with the consciousness of it. This curious complexity of composition can best be illustrated by a thought of oneself, since self is the most primitive subject of reflection. 'I desire' signifies that I am unified with the consciousness of a desire, and, as a matter of fact, my consciousness is the same as myself, and my feeling of desire does not exist for me apart from my consciousness of it. In this illustration thought is self-perceptive. In non-perceptive thought the elements of consciousness and feeling present themselves in idea, instead of in actuality. But the three elements are unified as ideas, because they are unified in self-perception.

The feeling which enters into this complex as an actuality or an idea, may be active or passive. It may be one of acting upon, or of active association with (that is to say, of possessing) another individual. Or it may be the experience of a psychic or mental state, such as that of an emotion, a sensation of being acted upon by, or of being passively associated with, another. In either case its bearings are signified by the connexion in relationship of traits, which may be elaborated into great complexity. Apart from the subject and verb, thought consists of traits which define or are ascribed to its subject. And, from one point of view, the verb may be regarded as a trait which is connected through consciousness with the subject. For, while it is true that the feeling which is unified with personality in the verb is *individualized*, it becomes a trait of consciousness, if personality be eliminated, as in the verbal participle.

Since the verb includes the element of consciousness it must comprise the element of *time*. For consciousness is a train of sequences—a succession of endings, continuings, and beginnings—of sensations that become recollections, of expectations that become sensations—a stream of changes in thought and feeling which flows unceasingly during our waking hours. Accordingly a state of consciousness must be in present, past, or future; the verb must be timed.

Oneself is one's personality, and we endow others with personality by projecting our personality into them—that is to say, by associating an idea of personality with them, as well as ideas of consciousness and of traits, and unifying the three when their appearances are such as would accompany the unification in ourselves. It is in this fashion that we ascribe conditions, not only to other persons, but to inanimate things. Thus we speak of a house as 'standing', of a road as 'running', and of a subject as 'stiff'. That is to say, we very generally interpret the world around us by ourselves. But we may of course derive a feeling from the observation of our environment. It is from purely sensory experience that we ascribe to a plant the condition of 'flowering'.

The verb is at its simplest in the concepts of 'being' and 'having', which have come into use as auxiliaries. 'Being' is, in fact, *consciousness in itself*, irrespective of any trait. It can then be used to signify the unification of a personality with the idea of an individual, a kind, or a trait-class. 'I am desirous' means that I am conscious of the unification of my personality with the class 'desirous'. 'Having' signifies *consciousness of association.*[4] 'I have a desire' signifies that I am unified with the consciousness of being associated with it; and by the use of this verb one individual or thing can be connected with another. The use of these primitive verbs as auxiliaries is drawn from our own experience. For, if I am suffering, I may think that 'I am ill' or 'I have an ailment' as well as that 'I ail'. *Being* can be used to express present, and *having* to express past time, because unification with a class is a feature of sensation, whereas the recollected past clings to us by association.

Individuals, kinds, and abstracts are expressed by nouns, in the nominative case when they are the subject of thought, in the oblique cases when they are used to form traits of relationship. A particular individual or

[4] The most obvious illustration of association with oneself is afforded by one's clothes, and it seems probable that 'having' originally meant 'wearing'. Following a Latin derivation from *habeo*, we speak of some clothes as 'habits'.

kind is denoted by the definite, one member of a kind by the indefinite article. Personal and relative pronouns express individuals, kinds, and abstracts by reference; the possessive pronouns signify that they are in associative relations. We regard other individuals as in the second or third person according as we are speaking to them, or thinking or speaking of them. But in addressing another, the third person, or a thing, is frequently substituted for the second person as in such an expression as 'your Excellency'. This is more deferential because it is less direct.

Thoughts and sentences owe their elaboration to the complexity of the traits[5] that are employed as attributes or complements to define or amplify the subject and the condition that is ascribed to it. These may be elaborated in such detail as to require separate dependent sentences for their expression. Traits of character are generally expressed by adjectives, or adjectival sentences, or are included as unified feelings in the verb—as for instance, 'the sea is rough', or it 'roughens'. Traits of relationship may be distinguished as *circumstantial* and *adverbial.* The former are expressed by the use of case-endings or prepositions which introduce the individuals or things that are brought into relation with one another. They may be included in the verb, as in '*out*running' or '*in*volving'. Through this relative connexion individualized ideas are drawn into service as traits: it is a trait of me that I am 'in the garden'. The accusative[6] expresses the cause or stimulus of an action; this is in relationship to it as its *object.* The genitive defines an individual or kind by its origin, or its accompaniments, that is to say, by its association in sequence or coincidence; to be proud *of* another is to be proud *in consequence* of him. The other cases, or prepositions, express defining traits of method or instrument, motive

[5] The Categories of Logic attempt to classify them.

[6] To accuse another of a thing is to assert that he is the *cause* of it (Skeat's *Etymological Dictionary*).

or purpose, and the association of others. They also express point and duration of time, point and direction in space.

A trait may have relationships of character with itself, that is to say, relationships of phase, degree, intensity, number, and manner. These traits of traits (or 'sub-traits' as they may be called) are generally expressed by adverbs, and may accordingly be termed *adverbial*. A state of consciousness can be isolated from the degree of its intensity or the manner in which it manifests itself; and, since the state is expressed by the verb, intensity and manner are signified by adverbs[7] or adverbial sentences. It is noticeable that adverbs are also used to define place and time; this leads one reflect how far place and time are, *in idea*, merely traits of consciousness.

It would be beyond the scope of this book to attempt a comprehensive examination of our various concepts: such an essay would attain the limits of a dictionary. But we may usefully review certain typical classes of them, in order at once to expand the foregoing conclusions and to test their reliability. It may be observed here that abstracts differ from traits only in being individualized, so that we may take one or the other indifferently as illustrations. And a feeling may be regarded as a trait of self as well as an individualized concept.

We have seen that the materials from which concepts are evolved may be either environal (sensory) impressions, or impressions of self, the latter including those of internal touch, of neural conditions, of psychic activity, and of consciousness. The simplest of traits are those derived from sensory experience, such as of surface, colour, form, sound, taste, and of number. Since, however, we receive simultaneous impressions through more than one sense, some of these trait-classes are

[7] But adverbs may of course be used to graduate traits that are expressed by adjectives—as in 'very good'.

more complicated than they appear to be: *round*, for instance, is derived from recollections of sight and touch, combined in association because they are connected by a familiarity of rhythm.

Ideas of psychic feelings are associated with ideas of sensory impressions when the feelings and impressions have occurred in rhythmic coincidence. A thing is *great* or *small*, according to the amount of energy which is required in order to compass it, and we compute the amount of this energy by dividing it into units of distance, as feet and inches. It is *near* or *distant* according to the amount of energy that is expended in reaching it, and this amount is similarly computed. A thing is five paces distant if it requires five paces to reach it. These traits are individualized as its *size* or *quantity*, *length*, *breadth*, and *depth*, and its *distance*. A thing is *heavy* or *light* according to the amount of energy that is needed to lift it, the derived abstracts being its *heaviness* or *lightness*. But its *weight* may also be regarded as an independent energy of resistance.[8] As such it is individualized, and yields the essential trait-class of *weighty*.

Our concepts of movement are in reality rapid sequences of ideas. Our impression of a step, a blow, or an utterance is that of a succession of internal touches with which effort is associated and an idea of purpose. In thought ideas of these touches follow one another as did the touches themselves in sensation, since the ideas have been associated by familiarity of rhythm. In fact, the operations of the brain repeat in idea those of the body. A progressive movement is one in a certain direction—that is to say, along a distance towards a place. Movement outside us is a sequence of changes in visual size and perspective, with which are associated ideas that have their origin in touch. For our appreciation of distance is, as we have seen, tactile. These changes in size and perspective accompany our own progressive

[8] The *inertia* of physical science.

movements, and we may be in doubt as to whether they result from our own progress, or from movement outside us, if no feelings of our own come into play. Consequently, when a train that is at a standstill close to us begins to move, we have the illusion that it is ourselves that have started. Our concepts of movement, then, include visual as well as tactile ideas. We can see the movements of our limbs, and accordingly associate impressions of sight with those of internal touch. Concepts of utterances are similarly complicated. For since words are heard—and seen if written or printed—ideas of sound and sight enter into our concepts of them.

A generalized idea of a movement is that of a *kind*—a 'blow' or a 'word' for instance. From the kind may be formed an essential trait-class, and from this again an abstract idea. *Active*, for instance, is a trait-class, from which arises the abstract idea of *activity*. Abstract ideas differ widely from kind-ideas, since in the process of abstraction concrete sensory elements drop out. Yet we commonly employ the same words for both, and obscure the difference between them. 'Movement' and 'utterance', for example, may express abstract ideas of moving and uttering, as well as ideas of concrete kinds of activity.

We have seen that psychic conditions in themselves are merely distinguished by their phase, as expansive or contractive, and by their intensity. They can, then, afford no definite ideas unless they are combined with ideas that are derived from sensory impressions or internal touches. We may, perhaps, have an indefinite consciousness of pleasure and displeasure as an idealess feeling of happiness and unhappiness. But, generally, these feelings are associated with ideas of stimuli: it is these which particularize them as definite pleasures or displeasures. We distinguish, as *good* and *bad*, things which, being favourable or unfavourable, excite pleasure or displeasure. We also term them 'pleasant' or 'unpleasant', confusing cause with consequence; so we speak of 'joyful' news, or a 'curious' thing, although the joy or curiosity is not in the news or thing, but in its effect upon us

Our concepts of emotion, whether active or passive, similarly include ideas of their stimuli. We cannot conceive of actual concrete love or anger without thinking of their causes, although from their kind-traits (loving and angry) we form abstract ideas of them—for which, however, we have no separate words.

Traits of emotion, action, and utterance are usually ascribed to ourselves and others by the use of verbs—very elaborate concepts which have already been analysed. The traits which they express are generally highly complex—that is to say, they include implications which are woven into a whole by familiarity of rhythm. In some primitive languages, indeed, a verb may include an idea of its object or cause: in Cherokee, for instance, there are distinct words for 'washing oneself' and 'washing clothes'. *Being* and *having* merely signify unification and association. But *seeing* implies sensations of sight, *thinking* the procession of ideas in a connected chain, *walking* ideas of movement of the legs, *going* and *doing* ideas of movement with a purpose, *giving* ideas of movement, of purpose, and of another person, *involving* the idea of a knot to which two strands contribute. *Sewing* implies the idea of a needle as instrument. *Rejoicing* and *grieving* involve ideas of gain or loss. 'I can' or 'I am able' is a unification of self with a feeling of doing that is followed by an idea of success.

It hardly needs to be explained that neuter verbs express feelings of doing or suffering in themselves; active and passive verbs feelings of doing and suffering caused by an external stimulus; reflective verbs feelings of which oneself is the cause. In the infinitive mood the verb is an idea of *purpose.* It takes the participial form, as a trait-class (*moving,* for instance), if the idea of personality be eliminated. But, since it retains the element of consciousness, it is in present, past, or future time.

From neural conditions on the physical plane are derived such ideas as those of hunger, alarm, relief, and satisfaction. Also those of success and failure, from which concepts of extraordinarily pervasive influence are formed. For we isolate from them the traits of power and superiority, weakness, and inferiority. Success is power when regarded as the consequence of resistance [*]: superiority when regarded as the cause of its effects. Ideas of superiority, that is to say of *excellence*, when associated with ideas of sensory impressions or feelings, give them the *aesthetic* quality that excites our admiration. When associated with other persons they affect us emotionally, exciting our respect or jealousy. And when associated with ourselves they arouse the exaltation of pride, because this revulsion follows success in neural sequence.

The nervous sequences of successful resistance followed by pride, and of failure followed by shame, are the origin of our concepts of what is *just* and *right*. For we expect to find this invariable succession in our environment—to discover that virtue is always rewarded and vice punished. Injustice is opposed to this expectation, and it therefore arouses the angry revulsion of *indignation*. And, since in nervous succession pride requites virtue, and shame vice, the sequence affords a relative concept of 'balance', which is expressed in our ideas of *worth*, *merit*, and *value*. There are degrees of worth, because there are degrees of intensity in resistance and in pride and shame. Another nervous sequence is that of a feeling of respect upon one of power: ideas of these feelings combine in sequence to yield concepts of *dignity*, if respect follows power, and of *honour*, if the sequence is reversed, and power follows respect.

From continuity and discontinuity in repeated coincidences and sequences there arise the relationships which

[*] Success or failure without resistance is good or bad luck.

are individualized as certainty, probability, possibility, peculiarity, and necessity. If a coincidence or sequence is invariable, it is certain; if variable, it may be either probable or possible. It is peculiar, if one of the traits that are in coincidence or sequence never occurs without the other: it therefore inevitably involves the other: there is no alternative, and this is our idea of *necessity*. These relationships are of essential importance in guiding the process of inference. They are ordinarily associated as adverbs with the trait which they define (always, sometimes, seldom, only), but may be used adjectively as in such a phrase as 'It is probable that'. It may be remarked here that, when continuity or discontinuity of coincidence occurs, not in repeated experiences, but in a single association of a number of individuals with a trait, or of a number of traits with an individual, it is expressed adjectively by 'all' or 'some'.

To review this complicated chapter. The ultimate materials of which our concepts are composed are the traits into which the brain disintegrates impression- or record-ideas, including those of internal touch and psychic activity and consciousness, as well as those received through the sense organs. Impression- and record-ideas are disintegrated, and their elements isolated, by discontinuity of occurrence, yielding traits that are either essential or accidental. These are generalized by being unified with identical traits similarly isolated from other record-ideas. Groups of essential traits are individualized if they are isolated from their environment, and include substance or energy or possess continuity. From individuals are formed concepts of kind, and from the latter essential trait-classes and trait-abstracts. Accidental traits are those which are variable or are shared by two or more kinds. They are generalized as accidental trait-classes and individualized as trait-abstracts.

Concepts of traits and abstracts generally include elements derived from feeling as well as sensation. They may be of character or of relationship according as they are formed by the generalization of experiences which are identical in themselves, or are identical in the fashion or manner of their association. The latter class comprises those which are termed 'circumstantial' and 'adverbial'. Ideas of traits may be formed of ideas in sequence as well as of ideas in coincidence, following one another, in the former case, because the impressions from which they are derived always occur in succession. Traits may be expressed at some length in dependent sentences.

The ascriptions or predications of thought are concepts of consciousness with which are unified ideas of personality and conditions of feeling. The feeling, so unified, is an individualized idea. But it may also be regarded as a trait of consciousness. Traits are employed as attributes to define the subject of thought, and, as complements, to expand the condition that is ascribed to the subject. Accordingly ideas of traits are not only the materials from which concepts are evolved: they form all the links in the chain of thought except the subject and the unification of the subject with consciousness which is the basis of the verb.

Our concepts of time, space, matter, and force are of such complexity and importance that they will be considered apart in the chapter following. But, even so, it will be felt that this is a very summary sketch of an exceedingly abstruse subject. Apparently, however, it suffices to show that the formation of concepts may be regarded as the outcome of nervous processes, if we grant that familiarity and strangeness of rhythm and trait are attractive and repulsive forces, and that record-ideas may be disintegrated into traits and the traits be drawn into other connexions. Of the conclusions to which it leads, perhaps the most striking is the immense importance of our *feelings* as materials of our concepts. We draw far more from ourselves than from the world outside us.

CHAPTER VI

TIME, SPACE, MATTER, AND FORCE

THESE four concepts are constantly referred to as the irreducible, or basic, elements of our knowledge of the outside world. They are certainly an invaluable means of reducing to a measurable form the consequences of the changes that occur around us. But they tell us nothing of the real nature of these changes or of their causes. For they are ideas; and like all ideas, are merely symbolic. The fact that we are impressed by outside conditions and changes proves that they actually exist and occur; and we can infer, from our impressions of them, that conditions are marked by real similarities and differences, that changes follow one another in variable or invariable sequences, and that they are often rhythmic, as is the rising and setting of the sun. But the ideas which we form of them are symbols and nothing more. Our concepts of Time, Space, Matter, and Force are symbolic; and it is, indeed, for this reason that they are so effective in explaining other symbols which represent—or misrepresent—our experiences.

Our concept of Time takes many forms. It may be that of a period (or duration); of a point of time; of past, present, or future time, and of an indefinite continuance. A period of time is evidently the interval between two changes or happenings; as, for instance, between a feeling of hunger and its recurrence, or between the rising and setting of the sun. We cannot think of a period that has not a beginning and ending. Failing an end, time becomes an indefinite continuity. We appreciate the length of a period by marking it off into subdivisions. It is subdivided by our sensations, our feelings, and our thoughts, and by the changes that occur outside us, as by sunrise and sunset and the phases of the moon. We may subdivide it more definitely by the use of equal rhythmic units, as by the minutes and hours of the clock. It is then clear that our concepts of periods of time are of a twofold character, derived in part from experiences of ourselves, and in part from experiences of our environment.

A point of time is obviously the commencement or end of a period. If we think of yesterday morning we take it as the beginning of a duration which ends with the present moment, the length of which is computed by recollections of the experiences that have occurred during its course, or, more exactly, by the clock. If we think of 1815 as the date of Waterloo, we take the beginning of the interval between Waterloo and the present, the length of which is definitely computed in years and months.

Past time is a tissue of recollections gained from our own experience and that of others. Present time is that of actual sensation : this is, indeed, the meaning of the word 'present'. In languages which have developed auxiliary verbs, the past is expressed by *having*, and the present by *being*, because our recollections are associated with us by past experience, while we are classified by the feelings of the present. Thus 'I have gone' means that 'I am associated with a recollection of going'; 'I am going' that 'I am classified as going'. Future time is the interval between an expectation and its fulfilment. We cannot think of the future apart from an expectation. This very frequently leads to a volition, and hence, in the verb, future time is commonly expressed by 'will'. But an expectation may also be based upon a possibility, probability, or certainty, upon the necessity which presses when no alternative is open, or upon the obligation which rests when the alternative is disadvantageous. Hence an idea of the future is implied in the words *can*, *may*, *shall*, *must*, or *ought*. For in all these cases the meaning of the phrase can be expressed as an expectation. 'He must go' means 'He is expected to go without fail'.

The formation of expectations is a process in which the brain is influenced by a motive that has its roots in instinctive life. For search is the inevitable consequence of an appetite, and in thought the cause of search is an expectation. The satisfaction of both an appetite and an expectation involves the pursuit of an object which is unperceived. Hence our concept of future time may ultimately be traced to the physical plane.

Now this analysis shows that time, as a concept, is an adverbial trait, or sub-trait, of consciousness, exactly as intensity is a sub-trait of psychic energy. Consciousness is the condition of the individual who is perceiving and thinking: time defines this condition and is, therefore, a trait of it. We can individualize it as a continuity, as we can individualize all traits—'probability' for example. But we commonly introduce it into thought by the use of adverbs, or prepositions, the form which sub-traits assume in grammar. Individualized as an abstract continuity, it gains consistency from the fact that we cannot imagine a rhythmic succession to which a beat cannot be added. We may infer that in this we are influenced by the beating of the heart, which has never begun and will never cease during conscious life. But this concept, based upon experiences of self, is out of accord with external experiences and may lead us to impossible conclusions. For infinite time possesses no duration, and is simply an indefinite repetition.

Let us now turn to Space. This is also a manifold concept. Its elementary meaning is that of a distance, ending, like a period of time, in two points—*here* and *there*. Now in discussing the origin of this concept, we must not regard sensations of sight as indispensable. For an idea of space is undoubtedly formed by the blind. The concept of distance must, then, be derived from the sense of touch, and the feelings which accompany it. These yield it as a feeling of movement in a straight line which is accompanied by energy and terminates at a goal. Distance is, in fact, the interval between starting and arriving: it is a period, like time, measured not in units of continuity, but in units of energy—as consisting, for instance, of so many *paces*. Accordingly it is a sub-trait of movement—that is to say, of velocity, and, as a concept, has no existence apart from ideas of movement. It is clear, then, that movement, not space, is the primary concept: distance and place define movement, and are therefore commonly expressed by adverbs and prepositions.

Movement is qualified by intensity or amount of energy, as well as by distance. Time enters into the idea of an amount of energy, since this depends not only upon its intensity but upon its duration. Accordingly distances are commonly computed in time, and time enters into our concept of space. Movement that does not cease at a terminal point is simply energy that is directed forwards—that is to say, velocity as a general concept. It is defined by science as a ratio between units of distance and units of time. But, since distance is merely a sub-trait of movement, velocity can be expressed more justly as a ratio between units of forward energy and units of time. So defined, it combines two elements which correspond with the two phases of nervous excitement—sensory and motor. For time does not exist apart from the current of ideas, and energy is motor excitement under another name.

Just as our concept of Time, as a period, arising from within ourselves, is fortified by our percepts of external changes, so our concept of Movement, primarily derived from our own activity, is extended by percepts of movements that occur outside us. We can perceive them through impressions of touch, although sight increases their definiteness. If we think of the stars as *fixed*, they appear to be studs set in a dome of crystal. But if we realize that they are moving towards or away from us, the dome is shattered : they are at different distances and the heavens are invested with Space.

We form fairly clear ideas of amounts of energy : we can estimate the degree of effort that is involved in a customary walk, or in lifting a familiar object. But it lacks precision, and distance, like time, is measured by being subdivided into equal units of length. Length is one of the traits of a surface, impressions of which ordinarily accompany forward movement. Accordingly we think of distance in terms of length, as we think of time in hours and minutes.

Impressions of sight *in themselves* give us no clue to distances. For near and distant objects merely appear large or small : their distance is associated with their appearance by recollections of the changes of appearance that accompany movement; and we shall see that, if ideas of sight are dissociated from those of movement (as may happen in cases of violent concussion of the brain), the visual impression of a scene is that of a flat surface of outlines and colours, which, lacking all distance, seem to be pressed against the eyeballs. Visual impressions are, in fact, as flat as pictures, and we endow them with distance, as we do pictures, by associating with them ideas that are derived from movement. Sight gives us impressions of transverse movements : but they are in miniature, and, apart from associated ideas of actual movement, merely tell us that moving changes occur in the world outside us—information which we also obtain through the sense of touch.

Distance implies direction. We obtain ideas of direction—other than that of moving along a surface—from the fact that, while the semicircular canals of the ear tell us that we are standing upright, by moving both arms we can represent all possible lines which run from the centre of a globe to its circumference. Accordingly we figure space as distance in all directions—as a sphere, the outside of which is the sky and the stars that are set in it. So conceived, space is a sphere formed of straight lines converging from its circumference to its centre. Our ideas of these lines are completed by the conception of straight rays of light, and if we consider these rays to be curved,[1] our notions of space are disorganized.

We may also think of space as an indefinite continuity, infinitely extensible and subdivisible, since we measure it by units of rhythm, and we cannot imagine a rhythm to which beats cannot be added, or the intervals of which cannot be rhythmically subdivided, as in a bar of music. But this concept, like that of infinite time, is not in accord with external experience and leads us to the impossible.[2]

We now pass to the consideration of Matter, or substance. Our concept of it is derived from touches which seem to meet with a resistance that is akin to our own, and is individualized as an energy. The weight of an object, as we have seen, is its resistance to being lifted: its solidity is its resistance to being compressed. Accordingly matter does not exist apart from resistance, and, in scientific conception, 'mass' merely expresses *inertia.* Mass in movement is conceived as inertia divided by velocity, for example, as a pound's weight travelling at the unit rate of 1,000 feet per second. It follows that inertia is mass multiplied by velocity, and this, of course, accords with experience. The more rapidly an object is moving the more difficult it is to deflect it from its course.

[1] This is the effect of the verification of Professor Einstein's conclusion that light is deflected by gravity. But this fact affects, of course, only our concept of space, not the actual condition that we term 'space', of which our concept is merely symbolic.

[2] As in the old logical puzzle of Achilles and the tortoise. The former runs twice as fast as the latter and gives it a hundred yards start. Unless we have a definite goal in mind, we can prove the untrue—that Achilles will never pass the tortoise: for when he has run a hundred yards, the tortoise will have run fifty: when he has run fifty yards, the tortoise will have run twenty-five; and, since, in this fashion, the interval between them can be divided to infinity, Achilles must always be behind.

The resistances with which we meet in the material world are bounded by surfaces, the principal of which is the ground upon which we live. An object which we can handle is bounded by surfaces which run into a continuous whole. We conceive of its substance as lying in three dimensions, because the vertical is our ordinary position, and from it straight, symmetrical, horizontal movements can be made sideways or fore and aft. Consequently we compute the size of flat surfaces by length and breadth, and of substances by height, length, and breadth, although these may give us very incorrect ideas of areas and bulks that are contained by curves. The outlines that are perceived in this fashion, with the assistance of sight, are the elements of our geometrical conceptions.

Force is plainly a concept that is derived from ourselves and is extended to objects that manifest resistance. In conceiving the forces that exist outside us, we are then arguing from ourselves to our environment. Our material surroundings undoubtedly possess some trait which gives rise to our idea that they resist us both passively and actively. But it may be something the real nature of which is quite different to that of psychic energy. We know that it varies in degree, and that it is followed by consequences, which, although only symbolized in idea, are actual changes. When we say that matter is expanded by heat we speak in a parable, for our ideas of heat and expansion are merely figurative. But when that condition which we term 'heat' is communicated to or withdrawn from that which we term 'matter', actual changes occur, with consequences that are to some extent measurable. Science, then, is concerned with actual conditions and changes, but establishes an acquaintance with them which is merely symbolic. We must not belittle it on this account. By the perception of similarities and differences, of impressions that are connected as causes and consequences, it is constantly extending symbolic understanding from the known to the unknown. Thus it concludes that light is vibratory, because the behaviour of light accords with that of rhythmic vibrations.

Reviewing these conclusions we perceive that, although impressions of our environment enter into our concepts of Time, Space, Matter, and Force, they derive their meaning from feelings of our own.[3] The concepts of Distance and Velocity are akin, both arising from impressions of forward movement, feelings of energy, and thoughts of time. They differ in that in distance energy is multiplied by time, whereas in velocity it is divided by time, because distance, being limited, absorbs a definite amount of energy, whereas in regard to velocity, time can only be used as a unit-rate.

Movement, Energy, and Time are then the most fundamental of our concepts; and it is remarkable that they should be derived respectively from the physical, the psychic, and the reflective plane. Movement is characteristic of instinctive life: indeed life, at its simplest, may be defined as a process for the conversion of external conditions into movement. Energy is psychic excitement. Time, as a duration, is an experience of intervals that occur in consciousness.

We may, in conclusion, refer rather more particularly to the origin of the processes by which we compute numbers, quantities, distances, and periods of time.

We compute by subdividing a whole into parts, or by constructing a whole out of parts. Both processes rest upon the identity of a whole with its parts, or of the parts of a whole with the whole, which may very possibly have been discovered by the manipulation of handfuls of pebbles,[4] as children play upon the beach. For ten pebbles are identical with three and seven pebbles lumped together, and may be subdivided into two parcels of three and seven: twenty pebbles are identical

[3] It follows that all scientific knowledge is ultimately based upon knowledge of ourselves—a science so little regarded as to have been left unnamed. It might perhaps be termed *Emautology*.

[4] Hence the word 'calculation'.

with four groups of five pebbles. These simple experiences would establish the first four rules of arithmetic.

Now the pebbles are not identical in themselves: some may be larger than others. But they are identical in representing units of rhythm, our susceptibility to which, as we have seen in chapter ii, descends to the physical plane, and is one of the elementary facts of our nature. Counting is the rhythmic subdivision of a group of units into a series of smaller groups, each of which exceeds its predecessor by one rhythmic unit. In their separateness the fingers are identical with countable objects, and accordingly counting can be effected by throwing them into successive groups. Numbers are primitively *quantities*: three represents a group of three fingers, five all the fingers of a hand—ideas which are embodied in the forms of the Roman numerals. The conception of a number as an individuality—such as is expressed by the Arabic numerals—appears to have arisen through the isolation and generalization of samenesses between different finger-groups. 'One' is common to one finger and two fingers, 'two' to two fingers and three fingers, and so on. This process gives us nine numerals only, and the series consequently stops at this number, 'ten' being formed by a combination.

When computation is applied, not to groups of units, but to concepts such as those of quantity, size, distance and time, these must be artificially subdivided into countable rhythmic units. One method of effecting this is to subdivide one object in terms of another, as when we say that a tumbler is as large as five wine-glasses. This comparison rests upon a sameness, for any two quantities are so far identical in that one can be expressed in terms of the other.[5] Another method of sub-

[5] From this identity emerges the idea of a relationship *in ratio* between two numbers or quantities, irrespective of their actual magnitudes. There are, of course, other numerical and quantitative relationships. One number, for instance, may be related to another in that the two are equal, or are contraries (*plus* and *minus*), or in that they contribute (as $\frac{1}{3}$ and $\frac{2}{3}$) to form a whole: $a-b$ is related to a in that the

division is to graduate into identical units which are arranged in scale—that is, in a rhythmic series. Thus distance is measured by being subdivided into lineal units of rhythm that are counted. Ideas of both rhythm and units are offered to us by our bodies—by the beating of the heart, and by the length of different portions of our limbs, as the first joint of the thumb (the inch), the foot, the cubit, and by the distance that separates two foot-prints.

Time, as a period, is measured by natural or artificial rhythms that occur in our environment. The alternation of day and night, the progress of the sun across the heavens, the changes of the moon, and the recurrence of the seasons, are all rhythmic. The day is conceived as containing twelve hours because it has identities with a year, and the year is subdivided into twelve portions by the changes of the moon. The actual division of the day into hours could be accomplished by the means of a rough sundial, such as we may imagine to have been set up on the plains of Babylon.[6] And if we realize that the primitive notion of rhythm is derived from the beating of the heart, we can understand why the hour is divided into sixty minutes, each of sixty seconds. For 60 is the square root of 3,600, and a pulse of 60 to the minute (or 3,600 to the hour) would be a not improbable average for an elderly Babylonian astronomer.

two differ only by the inclusion or exclusion of *b*. In algebra these various relationships in rhythm are used for the ascertainment of quantities.

[6] Ptolemy assigns to the Babylonians the invention of our subdivisions of the day, the hour, and the minute.

CHAPTER VII

PERCEPTION AND THOUGHT[1]

OUR perception of an object—instantaneous although it may appear—is a gradual process, in which the materials presented by sensation are interpreted and amplified by memory and intelligence, so that impressions of shape, colour, and change, which in themselves are merely kaleidoscopic, are elaborated into traits of character and relationship, and are individualized. Perception always involves the unification of an impression with a particular or general idea which already exists in the brain. It is indeed obvious that, if no idea of an impression existed, its recognition or classification would be impossible; and we may 'catch ourselves' in the process of unification when we hesitate in recognizing, or classifying, a person or object. The unification of an impression involves the association with it of other ideas which have been linked to the pre-existing idea by coincidence in sensory experience: the sight of a friend, for instance, recalls a coincident idea of his name. The associations thus introduced may be those of sequence, backwards or forwards, instead of coincidence—that is to say, they may have been linked to the pre-existing idea by having preceded it or followed it in sensory experience or in thought. So a feeling of chill recalls an idea of a preceding draught: a feeling of misconduct an idea of succeeding punishment[2]: the sight of one running

[1] 'Thought', like 'sensation', has an ambiguous signification, meaning either the *process* of thought or the particular thoughts which occur in the process. The *process* may be distinguished as 'reflection'. But, if we realize that 'thought' has two meanings, we may follow custom and use the word for the process.

[2] Since sequences are rhythmically connected *pairs*, their order is

an idea of haste. Associations in coincidence are part and parcel of perception : associations in sequence complete perception by the process called 'inference'. It is obvious that both are primarily memorial. Unless we recollected that fire burnt, or that a draught led to a cold, we should not associate these consequences with one or the other.

Accordingly, perception involves a unification that is accompanied and followed by associations. The process of association obscures the unification which attends or precedes it. But without unification it would be impossible. Mere sensation would clearly be unable to draw a recollection out of its retirement, unless the sensory impressions were unified with a pre-existing idea of an individual, a kind, or a trait, with which the recollection had previously been associated by familiarity of rhythm.

First, of associations in *coincidence*. A vast number of our ideas are linked with others by the fact that they have occurred together in sensory experience. There is demonstrative proof that our appreciation of the solidity of the objects that we see, arises from the intimate association of ideas of sight and of touch, since these may be dissociated by a violent shock to the nerves, with the consequence that visual impressions lose their substantiality. One of the most striking cases is that of a young American clergyman,[3] who suffered a severe

reversible, so that pain may recall an idea of misconduct. Thus a dog, whose paw has been accidentally trodden upon, thinks it has done wrong, and, after the first squeal of pain, sets itself to apologize. There are many good people to whom the misfortunes of others are proof positive of sin—Job's comforters, who forget the lesson of the 'Tower of Siloam'.

[3] Named Hanna. The case is recorded at length in *Multiple Personality* by Sidir and Godhart (Appleton, New York). 'The man, blind from birth, on whom Cheselden performed his historic operation, on first obtaining his sight, described the visual field as flat, and its contents as seeming to touch his eyes' (Schafer's *Text-book of Physiology*, ii. 944). 'Those who have been completely blind from birth, and have been operated upon late in life, have to learn by repeated experiences, especially of touch, to distinguish different objects in the field of vision from one another' (Hinshelwood's *Letter, Word, and Mind Blindness*, p. 2).

concussion in falling from his dog-cart, and completely lost all his memorial associations. He could recognize nothing, words meant nothing to him, and he was as helpless as an infant, since all co-ordinated capacity for uttering words or performing the simplest actions had been completely shattered. It was necessary to teach him afresh by many weeks of lessons and practice. When he had recovered ability to express himself, he described his sensations of sight, on first regaining consciousness, as of a flat mosaic of colours pressed against his eyeballs. Vision gave him no ideas of solidity and distance, and he was obliged to reassociate them by actually touching the objects which he saw.

As impressions of sight recall those of touch, so impressions of touch recall those of sight. The feeling of an object recalls its appearance. Ideas of sight and sound are similarly coupled. The sound of footsteps instantly reminds us of some one who is walking, and the noiseless appearance of footprints on the ground before us would be so alarming that it is an effective incident in a ghost-story.

We have already referred to the association of energy with objects, by which they are invested with weight and distance. We similarly connect particular emotions with certain appearances [4] of feature and with certain gestures and utterances, and invest other persons with these emotions if they manifest these external traits. So we perceive,[5] or *understand*, the mental and emotional states of others. They lie beyond the scope of direct sensation, but are so intimately associated with their

[4] Since feelings are not *invariably* manifested by facial expression or conduct, the two can be dissociated in coincidence, and their connexion be regarded as in sequence, backward or forward. We may *infer* that another is pleased because he is smiling, or that he is smiling because he is pleased.

[5] That is to say, we understand other persons through their expressions : there are no other 'windows of the soul'. We appreciate a cinema film because it represents a series of changing expressions, which are as effective in recalling ideas of feeling as if they were manifested by living individuals.

manifestations in expression, that we seem to 'divine' them. By means of these associations we understand a drama or a picture; and we endow animate objects with life because they manifest the movements that are linked with life in idea. We go further. We invest inanimate things with feeling if their appearances correspond with those that manifest certain feelings in ourselves. Thus we speak of a village as 'lying upon the hill-side', or of a church-spire as 'pointing to the sky'. Lying and pointing are obviously *feelings* which accompany certain positions and movements. And it may be observed here that, since in ourselves no change of conscious activity occurs without a cause—the feeling which precedes it in consciousness—we cannot think of a causeless change in Nature: if no cause is discoverable, we still figure one, as when we say '*It* rains'. The phenomena of Nature must have an originator or Creator, and, failing the idea of one, we regard Nature herself as an active agent, or individualize such processes as 'evolution'. We must think of surrounding activities anthropomorphically, for we can only interpret them by ourselves.

The words of another are in fact gestures [6] made with the vocal organs instead of the limbs. Words illustrate the coupling of ideas very strikingly. Our idea of the muscular utterance, sound, or sight of a word is linked in rhythmic coincidence to the idea which it represents. We learn a language by the formation of impression-ideas of its words, and establishing rhythmic familiarity between them and the ideas which they represent: and a language may be lost, either by the decay of the word-ideas, or the severance of their connexion with thing-ideas.

[6] The identity between words and gestures is illustrated by the curious fact that record-ideas of word-utterances are with right-handed persons stored in the left, and with left-handed persons, in the right hemisphere of the brain. (Owing to the crossing of the nerve-paths the left hemisphere is connected with the right, the right hemisphere with the left side of the body.)

Accordingly words are *emblems* or *symbols* of a certain kind—'idea-signs' which recall a meaning by memorial association. The use of idea-signs plays a part of immense importance in our lives. It introduces us to a complicated subject that will be discussed at a later stage. But we may observe here that, although these signs retain their meaning by memorial association, they may have originally derived it from connexions that have been established by familiarity of *trait* as contrasted with familiarity of *rhythm*. In the former case they are connected with their meaning by the sameness which underlies a resemblance or analogy: in the latter there is no such similarity or analogy between the sign and its meaning, and the connexion between them has simply been established by association in perceptive or reflective experience. We may distinguish the two classes respectively as *emblems* and *symbols*. A picture or statue is an emblem. Toys are generally emblematic. So are many of our manners; dignity is emblematized by a stately carriage. Souvenirs may be emblematic: such are locks of hair and photographs, since they possess an identity with the person whom they represent. But they may be merely symbolic, such as rings or keepsakes. Words, we shall see, may be of one class or the other.

Next of the associations in backward or forward sequence the use of which constitutes inference. They are memorial associations, which connect one idea with another, as its precedent or consequent. They may be derived from either perceptive or reflective experience, that is to say, through sensation or through inference: it is a sensory (or perceptive) experience that fire burns, a reflective experience that all men are mortal. The latter may appear to be rather a coincidence than a sequence. It is *actually* a coincidence, but presents itself in thought as a sequence, and we accordingly think that to die is a *consequence* of being a man. These sequences, when used in explorative or expectative thought (of which more hereafter) enable us to divine the past or to predict the future. We know, for instance, that water wears substances away. Accordingly we may infer that pebbles which are rounded *have been* worn by water, or that a water channel *will not* last, unless it is made of hard material.

The value of associations in sequence as a means of reconstructing the past, or of foreseeing the future, depends, however, upon their generality or peculiarity, their variability or invariability. These, we have seen, are adverbial sub-traits. Ideas of them are derived from experience, and are recollected along with the associations which they qualify. An expectation of rain from a fall in the barometer is only *probable*, because rain is not its invariable consequence. We infer with probability that a man who is wearing a tall hat on Sunday morning has been to church—almost with certainty, if he is carrying a prayer-book, since this is peculiar to church-going, and hardly leaves a possible alternative: he *must* have been to church.

Associations in sequence may be initiated by the idea of an individual or by that of a happening that is unattached to a cause, as, for instance, the disappearance of one's umbrella. In the first case association may run from the kind of the individual to a trait, from a trait to kind, or from one trait of character or relationship to another. Hence we may infer that a predaceous animal has large canine teeth, or that an animal with large canine teeth is predaceous, that a thin-toned piano has been long in use, and that, in this case, its action must be antiquated. A trait of relationship, as we have seen, may be of circumstance, time, place, purpose, manner, or intensity. I argue from one trait of relationship, through another, to a third in inferring that 'I must walk quickly, because the train leaves in a quarter of an hour and it is a mile to the station'. When inference is initiated by a happening, its course runs from traits of relationship through a sequence to an individual. We infer, for instance, that an umbrella has been taken by a particular caller, because it was in its place an hour ago, and could only have been taken by one who had called since.

An inference as set out completely in consciousness consists of two reasons and a conclusion. The reason [7] that generally occurs first is one of observed fact; the second, which is the association in sequence, is one of perceptive or reflective experience. The conclusion incorporates in the sequence the subject or trait which initiated the course of inference. It is the idea-connexion which it was the object of the inference to secure. We are conscious of these three stages when we infer under the pressure of doubt or hesitation—that is to say, when inference is 'reasoning' or resistant. But ordinary inference is so rapid as to elude consciousness. The conclusion commonly presents itself before we have appreciated the reasons which led to it: they occur to us after the conclusion; and hence, in explaining a conclusion, we generally state the reasons after it, introduced by a 'because' [8] instead of a 'therefore'. And the reason of experience is, as a rule, omitted altogether. We say that 'he must be ill because he looks pale', without explaining that paleness is ordinarily the consequence of illness. This is left to be understood.

The process of reasoning inference will be considered and illustrated in the chapters which follow, and with this passing notice of it we may transfer our attention to the perception of differences and changes. The

[7] The reasons of fact and of experience are the Minor and Major premisses of Logic. In some cases, as we shall see, the Major precedes the Minor premiss. But there must be a fact to introduce it into thought, unless we are arguing from a hypothesis or to convince others.

[8] 'Because' introduces a precedent, as the *cause* of a consequent, and 'since' a consequent which has become a precedent. But there is precedence in reasoning as well as in actual experience, and hence 'because' is used ambiguously. In such a phrase as 'he is on crutches, because he has had an accident' it refers to actual experience. But, if we use a sequence of reflection, we may argue that 'he has had an accident, because he is on crutches'. This ambiguity is the source of much confusion of thought.

impressions which we receive do not unify when they arise from vibrations of different qualities, or are successive in time. For the former of these reasons the various features of a scene, as impressed upon the retina, stand apart from one another as different objects, although, if they have elements in common, they may be *like* one another. If all the impressions are identical the scene is uniform, as in the appearance of a cloudless sky. Successive impressions of the same unchanged scene are distinguished from one another because they are successive in time although identical in feature. When the successive impressions are different, the variations are perceived as changes, incidents, or happenings, which either have occurred or are in course of occurrence. In the former case the difference is brought out by unifying the present scene with that which is recollected from the past. Thus we notice that the weather has changed during the night by unifying to-day with yesterday and isolating the difference. An incident in course of occurrence is a sequence of changes, which involve ideas of movement in ourselves or outside us, if they are associated with ideas of touch.

When the difference is not outside us but within us—when it is between, not impressions, but the ideas with which impressions are unified or associated—the condition of confusion comes about which we term 'doubt' or 'hesitation'. We may, for instance, identify a star with the polestar, and then discover, 'to our confusion', that its situation does not accord with that of the polestar; or we may hesitate whether we shall associate ourselves with a game of golf or a tea-party. This state of perturbation is, as we have seen in chapter iii, ended by the resistant effort which is *curiosity* when it is stimulated by the unknown, *choice* when it is led by competing ideas of consequences. Or it may be followed by the expression of curiosity in speech, as a question. Accordingly differences control perception by perturbing the brain; so they 'fend off' incorrect unifications, as we realize when we have much difficulty in recognizing an object. If differences are imperfectly appreciated, the identification of an object may wander far afield. This is so in the hypnotic state—and to a less degree when we are in an imaginative condition and differences are masked by striking similarities.

We may infer that the confusion of doubt or hesitation is, in fact, a disturbed condition of the current of cerebral motor excitement which flows from idea to idea during perception and thought. For it may reverse its phase, exactly as psychic excitement may be reversed from an expansive to a contractive condition. We may, moreover, infer that the reversal may affect the connecting nerve fibril, or the idea that is brought into connexion by it. In the former case there results a *negativing*, in the latter a *contrarifying* reversal. The connexion of ideas which has been made is maintained, but its phase is changed. This comes about in perception when a unification or association is contradicted by further observation or inference, or by the information that is given by another, which presents differences that invalidate it. Thus we may decide that our star 'is-not polestar' or 'is some other star'—the former negativing, the latter contrarifying the connexion. An affirmative and its contrary are, accordingly, two phases of a single thought, and we can understand why contraries so frequently suggest one another. From negative and contrarified thoughts ideas arise that retain these forms—*unsatisfied* and *dissatisfied*, for instance—which can enter into thought as if they were positives, involving no reversal of cerebral excitement. But contraries are often confused with negatives: we use, for instance, the word 'undress' instead of 'disdress' (*déshabiller*). And the contraries used in algebra are called negatives —to the obstruction of understanding.

Reversals may also be caused by a volition impelled by a like or dislike, as when we reject the suggestion of a walk and say 'I will not go'. The idea-connexion between ourselves and a walk has been established by the words of him who suggested it. We maintain the connexion but reverse it. In this case, however, the reversal is not in regard to the nature of a thing, but to a pleasure or displeasure which we should pursue or avoid. It is practical, not theoretical. This distinction constitutes the difference between two of the principal phases of thought—expectative and explorative. We may also negative or contrarify a suggestion out of ill-temper or contrariety. This illustrates the effect of emotion upon thought—a subject upon which we shall have more to say hereafter.

Now it is evident that perception involves thought, or reflection. It is, in fact, perceptive reflection, and the conclusion presents itself that thought—or non-perceptive reflection—differs from perception only in that it is independent of sensation, and uses recollections and ideas as its materials, instead of impressions of the senses. Indeed, it can hardly be termed independent of sensation, since the current of motor excitement which illumines its idea-connexions and renders them conscious is derived from sensation. We begin to think on awaking because sensation recommences. But connexions may be established when this current is not running—in other words, thought may be subconscious. There is no one who has not experienced the subconscious offering of a recollection which he has been consciously endeavouring to recall in vain; and it is a not uncommon experience that, on going to sleep upon an unsettled problem, one awakens to find it solved. Subeonscious thought is strikingly illustrated by the phenomena of hypnotism. By fascination—that is to say sensation so intense as to be paralysing—the cerebral motor current is arrested, and the hypnotic trance is produced. In this state ideas that are presented by suggestion are immediately accepted since, there being no motor excitement, reversals cannot occur so that differences lose their effect. So one under hypnotic influence will unify impressions on superficial samenesses, taking water for wine, a walking-stick for a snake, at the hypnotizer's bidding. And in the process of subconscious thought recollections may occur of impressions that struck the brain subconsciously, of whose existence one was unaware.

It is, then, clear that the process of thought is not dependent upon conscious control—that the attractive force of familiarity of rhythm and trait suffices to make the idea-connexions of thought, and combine them in a series. The idea-connexions may be merely repetitive, or ' retrospective ' : this is so when we simply remember perceptive or reflective experiences. They may be imaginative, taking a peculiar concrete form, which will be elucidated as we proceed. In other cases they consist, in great measure, of exceedingly rapid inferences, which introduce sequences drawn from perceptive or reflective experience, as is evident from the fact that most ascriptions of thought can be amplified by a *reason*, introduced by a ' because '.

Each idea-connexion, or thought, however complicated, and each dependent thought which is included in and contributes to it, ascribes a condition to the individualized concept that is its subject by unifying an idea of the subject's personality with an idea of consciousness—that of *being*, of *having*, or any of the active or passive emotional or mental experiences that are signified by a verb. The subject and condition are elaborated by the use of traits which, when of character, are generally expressed by adjectives; when circumstantial or adverbial, by prepositions that introduce individualized ideas into relationship, or by adverbs. These traits may be expanded into dependent thoughts —adjectival, circumstantial, or adverbial. Each complete thought is linked to its successor by the unifications or associations which we are about to consider. In a grammatical sentence these connexions are set out in a regular order. But in thought they may obviously present themselves in much simpler forms than those which are employed in verbal expression Indeed grammatical form is in great measure conventional: 'I have money' is expressed in Arabic as 'with me (is) money'. And the crude, ungrammatical expression of a string of ideas may be quite intelligible.

The process of thought is so difficult to analyse because it is veiled from us by the illusion that we think in words, not in ideas. Words are sounds made by muscular movements, ideas of which have become so closely associated in rhythmic coincidence with the ideas which they represent that the one seems inseparable from the other. But we can, and do, think without words, as do birds and beasts. We may arrive at a conclusion and still be unable to express it in words, until we have successfully hunted for an appropriate expression. We may hesitate as to the phrases in which to express a course of idea-connexions, and 'pick and choose' our words. That 'thing-ideas', not 'word-ideas', are really the material of thought is evident from the fact that two men, using different languages, may arrive at the same conclusion, if they think upon the same subject.

But, although (simply memorial reflection apart) we do not think in words, our ideas are generally accompanied by ideas of the words that symbolize them. If we liken the thinking brain to an electric chandelier composed of myriads of little lamps, some of which, now here, now there, are constantly breaking out into light and becoming extinguished, we must remember that each light may be double, since it may represent not only an idea, but the word which is associated with the idea. As we have seen, however, the two may be divorced. We may think in ideas without word-ideas as well as in ideas to which word-ideas are attached. We can *remember* a string of words that are quite meaningless,[9] as well as a string of words and

[9] Indian children will learn to repeat chapter after chapter of the Koran, although they do not understand a word of Arabic.

ideas. In thinking, there may be conflict between ideas and words, since the former come in a natural, the latter in a more or less artificial order, which is peculiar to each language and is forced upon us by memory. Accordingly, if we think in wordless ideas, the sequence, when expressed in words, must be readjusted by memorial phrasing; and, if we think in ideas with words, the association of words is liable to twist the sequence of ideas out of its natural channel. In fact, language, embellished as it has been by Art, may not only obscure but deflect the course of the ideas which it symbolizes, and we must disregard its ornamentations if we would penetrate to its real significance. If we would survey a cathedral from the engineering point of view, we must not lose ourselves in the appreciation of its decorative features.

There is a further complication which goes to render the analysis of thought so exceedingly difficult. We conceive that during thought perception is suspended. But, as a matter of fact, thought is constantly interrupted by self-perception—that is to say, by the intrusion of feelings, which appear to be thoughts. If a thought stimulates a psychic change, a percept of our condition breaks into our reflection as a trait of self, or a unification of self with it, as 'I think', 'I desire'. An emotion stimulated by a thought may give us a new purpose, switching the course of reflection into another channel. So also may the arising of an appetite or an external sensation. Our meditations may be interrupted by a pang of hunger as directly as by the luncheon bell, or the call of another.

In so far as reflection is inferential, the thoughts that enter into it are obviously connected with one another as the links of a chain of reasoning, although their connexion in sequence may be masked by the omission of some links from consciousness. For instance, in the sentence just written, it is *inferred* that thoughts are linked with one another, because they follow one another, and in experience things do not come to pass in a series unless they are connected. But we omit this reason because its application is 'obvious'. The *hypothetical* thoughts which figure so largely in reflection are inferential: they draw inferences from possibilities. There are, of course, other phases of interconnexion. One thought may be merely complementary to its predecessor, expanding it in detail, as is the case with relative and adverbial sentences. One thought may restrict the implications of another. But all these connexions are phases of unification or association, or of unification or association that is reversed.

Through *unification* thoughts are brought forward that explain, expand, or illustrate their predecessor, and are therefore united with it by a sameness. Two thoughts may also be coupled together conjunctively (and) or alternatively [10] (or), because they are alike and one suggests the other. *Association* introduces thoughts that have been united in sensory or reflective sequence. It may also connect two thoughts conjunctively [11] or alternatively because they are connected in rhythm. One thought may introduce another by its implications, which are shown by inference to need restriction. Their extension is limited by a thought that is introduced by a 'but' or 'however', or by prefixing 'notwithstanding' or 'although' to the first of the two. Hence 'although he was an Athenian, he was taciturn', negatives the inference which would ordinarily be drawn from his nationality. In the course of thought an alternative or possibility (introduced by 'if' when positive, by 'unless' when negative) frequently presents itself as the consequence of an inference: its consequence is inferred and is stated as a hypothetical conclusion. The reasons for an inference may, as we have seen, precede

[10] Hence 'or' may be used to introduce a pleonastic repetition, as 'in trouble or adversity'.

[11] Two individuals may also be connected conjunctively or alternatively with one trait, or two traits with one individual, either because they are alike, or because they are coincident in rhythm.

or follow its conclusion: in the former case 'therefore' is prefixed to the conclusion, in the latter case 'because' is prefixed to the reasons.

This illustrates the reversing of a sequence which, as we have seen, is a common occurrence in nervous life. A thought is a sequence, and we may therefore reverse the order of its arrangement: we may think that 'Diana is great' or that 'Great is Diana'. The order in which two thoughts are paired conjunctively, alternatively, or restrictively is always reversible.

The particular unifications, associations, and inferences which constitute a train of thought are determined by the *purpose* of our reflection. This appears to guide its course exactly as attention concentrates perception—by excluding ideas which are not related to the subject. The purpose of a course of reflection may be simply to explore the ideas which are its subjects, in which case it closely resembles perception. It may affect a practical issue—that of the conduct or speech, by which we may pursue a pleasure or avoid a displeasure. In either case the intensity of our purpose—the concentration of our thoughts—will depend upon our emotional susceptibility to the subject of reflection. We all know how difficult it is to think continuously upon an uninteresting subject. But susceptibility can be sharpened by practice, and in thinking, as in perceiving, the interest or attractiveness of a subject may be increased by persistent attention to it.

Courses of thought are of such protean diversity that one may well despair of classifying their phases with any approach to simplicity. Yet this problem will lose much of its difficulty if we realize that the functioning of the brain is influenced by motives which arise from the physical and psychic planes. In its external relations our physical life is concerned with movements of pursuit or avoidance. Thought which directs these movements is of the practical kind known as 'common sense', and, since in conscious life pursuit and avoidance are stimulated by pleasurable and displeasurable expectations, practical reflection may be distinguished as *expectative.* It may be swayed by an appetite, an emotion, or a desire that ends in a volition.

The features that characterize the psychic plane are active resistance and the development of passive conditions of pleasure and displeasure. Resistance that opposes itself to the disconcerting mental conditions of doubt and hesitation takes the form of curiosity—the reflective counterpart of courage—a propensity to add to experience by the assimilation of new ideas, that is to say, by connecting them with those which form the web of existing knowledge, instead of recoiling from them as unfamiliar. Thought which is urged by curiosity explores the ideas that present themselves, developing them after the fashion in which the eye apprehends the features of a landscape by travelling across it. We may accordingly distinguish it as *explorative.* It is, perhaps, the commonest phase of reflection—ranging from the purely trivial to the abstruse or intellectual—and expresses itself in action by inquiry, travel, and experiment.

Passive feelings of pleasure and displeasure stimulate thought which criticizes or *appreciates,* favourably or unfavourably, the ideas by which they are excited. We term these feelings 'passive', but all nervous excitement produces active consequences in some form or another, and appreciative thought is led by active likes and dislikes which express themselves in speech as praise and blame. They are conditions of attraction and repulsion, and evidently owe their activity to the influence of the physical impulses of approach and recoil, working in concert with psychic excitement and the brain. When a like or dislike is aroused by a psychic stimulus, involving an idea of superiority or inferiority in any form, it becomes admiration or contempt.

Moreover the energy of psychic excitement does not exhaust itself in producing these mental forms of approach and recoil. It may stimulate a further remarkable activity in the process of imagination—a creative mental condition of very peculiar character that exercises an extraordinary influence upon thought and action. The energy of pleasurable or displeasurable excitement, when acute, must liberate itself in muscular action: in unreflective life it produces the manifestations that are termed 'expressions of emotion'. When this liberation is effected through the brain, it involves, as its prelude, the formation of concepts that can be materialized by conduct or speech—the creation of images which can be expressed in a tangible form. This is *imaginative* thought which, it must be realized, is merely preliminary to imaginative activity in deed or word.

It follows that appreciative and imaginative thought are stimulated by the physical and psychic conditions that are felt as pleasure and displeasure. The latter is developed from the former. We are well aware that imagination involves appreciation—is, in fact, the fruitful consequence of appreciative thought.

Accordingly by physical and psychic promptings thought is impelled through the four channels which we have termed expectative, explorative, appreciative, and imaginative. Apart from the simple process of recollecting, the brain appears to have no initiative of its own. It is an instrument, not an energizer, and can, therefore, recover from injuries which would be fatal did they affect the spinal cord or the cerebellum. When untouched by physical or psychical influences its operations are, for the most part, simply reminiscent, that is to say, repetitive.

Reminiscence, we feel, is hardly a phase of thought. It may repeat in idea a succession of sensations or feelings, or a previous train of thoughts that have been formed in reading or listening to another. It may repeat these in the very words of our author or informant, and hence one who is endowed with a good memory has at his disposal a vast stock of phrases and periods, which may provide him with a flow of language that may be eloquent as well as copious if he has gathered them from artistic models.

The repetitive sequences of memory may be broken by a *forgetting*. An idea, or a word, fails us, and we are aware of its non-appearance because it leaves an association incomplete, with a consequent mental condition of perturbation. The 'switching' of the connecting fibril or nerve-path is impaired by disuse, and needs [12] time to adjust itself. The missing idea or word may be brought into connexion in 'roundabout' fashion, as when we recall the name of a friend by going through the alphabet for its initial letter; and, no doubt, when it suddenly presents itself out of the subconscious, it has re-established its connexion in this manner. It is remarkable that the susceptibility to rhythmic coincidence and sequence, which is the basis of memory, decreases so greatly with advancing years. In childhood memorial associations—muscular as well as mental—may be established by single casual experiences, whereas, later in life [13] many repetitions are required in order to bring about an association that will endure. But a single experience will suffice if accompanied by strong emotion—that is to say, if psychic feeling enters into the coincidence or sequence. A poignant emotional experience—such as a declaration of love, or an anguished parting—comes back to us, not alone, but mounted, as on a theatre stage, in a scene of elaborate detail. Hence the use of punishment as an aid to memory: a schoolboy who has been caned for a false quantity does not easily forget his error.

[12] Memory also depends greatly upon an adequate circulation of blood in the cortex. Sir Lauder Brunton has related that on one occasion, when excessively fatigued, ideas for an article would only come to him when he laid his head on the table; and many find that they think most actively when lying down.

[13] The fact that this applies also to the acquisition of muscular dexterities proves that these are also retained memorially.

For the sake of simplicity we have not, as yet, referred to the processes of perception and thought by which we acquire ideas from others, or compose thoughts for being communicated to others—that is to say, to the processes of comprehension and suggestion. A number of desires urgently press us to communicate with our fellows. A purpose can be most easily achieved, or curiosity satisfied, when one is assisted by another. We may communicate with others to assert our dignity or excellence, to disclaim or conceal inferiority, to help another or seek his aid, to oppose another or challenge his antagonism, or to protect oneself from his rivalry or opposition—the principal motive of deceit. Hence suggestive reflection and comprehension are of such vast importance in our lives.

Thought is typically *suggestive* when it is designed to influence others, to induce them to 'think with us', or act as we desire. We may influence them through sympathy, authority, or persuasion. In the first case our thought must be guided by an appreciation of the other's feelings: ordinary conversation is sympathetic, and, should it cease to be so, it no longer gives pleasure. In the second case we trust to the respect with which he regards us: we influence him by command. In the third case we rely upon the inducements that we can offer him, or the pleasure or displeasure, hopes or fears, that we can arouse. Suggestion may be made through acts as well as through words: the payment of money is a most effectual means of persuasion. Briefly then, suggestive thought follows a peculiar course dictated by a special motive—the desire to influence another.

Suggestive perception, that is to say, *comprehension* relies upon the memorial associations of words with ideas. The words which we hear or read are symbols to which ideas are attached. We fancy that we are 'in communication' with another, that thoughts are passing from him to us. But, in fact, the position is that of two telegraphists who by the use of conventional symbols are able each to call up ideas that are in the other's mind. Suggestion is, in fact, recall, and cannot act unless some elementary ideas have been acquired by perception. But it forms the ideas so acquired into thoughts and courses of thought, which we preserve in memory as sequences. So one can be taught that Australia is an island, if sensation and feeling have given him ideas of land, of water, and of 'surrounding'.

If thinking is a process which follows a definite course, determined by regular forces, why, it may be asked, do not all men think alike on a given subject? In the first place because their knowledge—their record of perceptive and reflective sequences—is not equally copious. Secondly, because individuals are unequally affected by prejudice and faith—emotional influences which, we shall see, control very powerfully the processes of inference. Thirdly, because the imaginative spirit is much stronger in some men than in others. And, fourthly, because susceptibility to familiarity of rhythm and of trait varies very greatly. Some men enjoy good, others suffer from bad memories. And, very commonly, the excellence or deficiency is not general, but affects particular classes of ideas. For it is common experience that one who has a bad memory for certain kinds of ideas can easily recollect others. This differentiation in associative capacity is strikingly illustrated by the phenomenal memory for figures with which some men are naturally endowed.

Susceptibility to familiarity of trait also varies from one person to another. This is the difference between the intelligent and the stupid. And here again we may notice that susceptibility may vary according to the *kind* of trait. Thus the mathematician is peculiarly responsive to traits of relationship in number and quantity, the inventor to those in mechanics, just as the musician is gifted with a special appreciation of tones and harmonies which enables him to link or combine them artistically When acute intelligence is combined with a retentive memory, and is assisted by vivid imagination, by appreciation of beauty, and by skill in expression, we have genius.

The principal points that we have endeavoured to make in this chapter are that when perception involves more than simple recognition, it includes the association of traits either by memory or by inference: these associations are of traits with an individual—idea-connexions that are, in fact, thoughts: inferential association proceeds with extreme rapidity except when it is interrupted by doubt; it is then a slower process involving two conscious thoughts that, as reasons, lead to a conclusion: when differences occur which impede unification or association, a nervous state of doubt or hesitation occurs that may end in the reversal of a connexion into a negative, or of an idea into its contrary: thought, apart from perception, consists of a series of unifications, and associations made, or reversed, that are broken from time to time by the self-perception of feeling: in thinking, the brain is influenced by impulses that reach it from the physical and psychic planes, rendering the course of thought expectative, explorative, appreciative, or imaginative: communication with another through words or gestures involves suggestive thought in him who makes and symbolic sensation in him who receives the communication, the latter involving the recall of ideas by the words that are received.

CHAPTER VIII

EXPECTATIVE AND EXPLORATIVE THOUGHT

THE *Expectative* or 'practical' problems of life are—What to pursue or avoid? and, How to pursue or avoid it?—that is to say, they are concerned with objects, and with methods or instruments. These questions are primitively decided for an animal by its instincts—by impulses that arise from the neural states on the physical plane of which we are conscious as the appetites, the shock of strangeness, injury, or difficulty, and the relief of familiarity, benefit, or help. And even in conscious life their instinctive imperiousness survives. One who is impassioned by love or anger acts or speaks 'thoughtlessly' or 'impulsively'—that is to say, without reflection. He acts *because* he is impelled to act, and for no reason derived from experience. So we may buy a thing against our judgement—against the counsels of the past—because we like it: a mother may 'spoil' her children by injudicious treatment, because she loves them: in pressing a defeated antagonist we may forget the future consequence of his revenge, because we hate him. In these cases the cause of our action is not a reason, but an emotion.

By the influence of faith we are also swayed impulsively. Faith is a condition that is stimulated by an idea of protecting or advantaging power associated with a person, an institution, or an idea. It is an alloy of respect and confidence, and has its roots in neural reactions and conditions on the physical plane that are spiritualized by psychic energy, and etherialized in consciousness. If we are loyal to a leader we unreflectingly accept his directions as to what we should pursue or avoid. We may have a similar regard for the laws or for current morality, since they are powerful and protective influences. It is true, of course, that the observance of moral rules may be the result of expectative thought—of calculated ideas of consequences. But it may also be forced upon us by unreasoning faith.

We cannot describe impulsive conduct as 'voluntary', since no willing comes into play. The will has involved from the instinctive effort to 'do something' when confronted by a difficulty. The course of this effort is directed by the instinctive propensity to approach the favourable and recoil from the unfavourable. This is extended from the present to the future—that is to say, approach becomes pursuit and recoil avoidance—by the influence of *expectation*, by the propensity to 'look ahead' for which we can find an origin in the instinctive promptings of the appetites. For these are peculiar in that they urge us to pursue or avoid things which are beyond the range of present perception. Expectation, as we have seen, is the foundation of our concept of the future—a possibility which forsakes us only in the hour of death. When our conduct is stimulated by expectations, possible pleasures[1] and displeasures suggest themselves in bewildering multiplicity. For each idea brings with it a train of *consequences*.

That these consequences will follow is inferred from the fact that they have followed their causes in the past, and are, therefore, associated with them, memorially, in sequence. If I am inclined to strike another, the idea of my action is unified with one of violence in general, which is associated as the antecedent of a retaliation; and the retaliation will be certain, probable, or possible according as it has followed invariably or variably. The chances of retaliation enter, therefore, into my consideration of consequences, and renders it 'calculating'.

[1] We shall see that the pleasure may be that afforded by an ideal. But this is for later consideration.

The appraisement of an act by its consequences introduces the alternative possibilities of doing it or not doing it. This affords the process of inference a hypothetical, or conditional, starting-point; and it proceeds to infer the consequences of the alternative possibility by connecting it with sequences into which it has entered in the past. Thus if, when walking to the railway station, I think that I shall be late, inference suggests running, and hypothetical inference suggests that if I run I shall get hot. By hypothesis we take into consideration the consequences of conclusions as to possibilities.

The conflict between alternative inducements brings about the mental confusion of hesitation. This is controlled by the resistance which is involved in steady deliberation, and ends in an effort of selection or choice. The effort is undirected by instinct, and is, therefore, amenable to the influence of pleasurable or displeasurable expectations. These are weighed, one against the other, in the process of deliberation, and, finally, we pursue the most attractive or avoid the most repulsive. We suppose that our choice is 'free'—and it actually contains elements of freedom, in that, as it arises, the effort to do instead of to suffer is undirected, and in that the choice of one temptation involves the resistance of others. But the course into which the effort is finally guided by choice is determined by the pleasurableness or displeasurableness of its consequences. One cannot claim spontaneity if, when deliberating over the gift of a subscription to a charity, he suddenly yields before the thought that his contribution will be published in *The Times.* This illustration introduces a fact of great importance. The pleasure or displeasure that influences choice may be psychic as well as sensory. It may be pride or shame in one of their many associations. Thus we may 'choose' to attend church because this is respectable, and, being so, gives us a feeling of self-satisfaction. And one who has fear in his heart may choose to stand fast, for fear of the shame of self-reproach, or the reproaches of his comrades.

A resolution is evidently an anticipated choice. The occasion of choice presents itself as an expectation, and we decide in advance. In fact we choose in present, and resolve in future time. Our resolves often fail us. Nevertheless it is the power of forming resolutions that endows man with the stability of purpose which distinguishes him from the most intelligent of dogs, and enables him to elaborate conduct as persistent and as complicated as the instinctive operations of insects. Our resolutions are beacons, lit by ourselves, as steering marks through life's temptations. We keep our eyes upon them. They set a course which seems to lead to a haven. We may swerve from it. But it gives a purpose to life, which would otherwise appear to be mere objectless drifting.

Expectations are based upon experience, and since experience is the foundation of reasoning, expectant volition is guided by reason. It is influenced in its pursuit of the pleasurable and avoidance of the displeasurable by sequences that have been impressed upon us by experience or by the teaching of others. So enthusiasm may be restrained by thoughts of 'safety first', and lust be held in check by fear of disease. But volition may be irrational, or emotional. This is so when it is influenced, not by expectations, but by hopes or fears, begotten, not of reason, but of likes and dislikes, reinforced by emotion. Hopefulness, as we know, manifests an enthusiastic or sanguine disposition: in despair one sinks into the profundity of nervous depression. Hopes and fears may lead to success or failure by inspiring one with resistant energy or by relaxing the strings of effort. But they are, in themselves, delusive —lode-stars or bale-fires which guide individuals and nations to erratic vicissitudes, not to steady progress.

Passing now to methods, or instruments, of action, it is clear that these are also primitively dictated by instinct, and that, when instinct loses its detailed control over conduct, *imitation* anticipates thought, as a means of shaping conduct to the needs of the moment. Imitation is the expression in action of ideas of action or utterance that are obtained from another. There are many facts to show that it may be involuntary: smiling, frowning, and yawning are, as we say, 'infectious': one who associates with a stutterer may catch the habit of stuttering. A more serious illustration is the extraordinary contagiousness of mob panic, or mob extravagance of any sort. Imitation dominates what is called the 'psychology of the crowd'. We imitate another when we act in accordance with his words as well as when we follow his gestures. For in both cases we express ideas that we receive from him. Hence example is a phase of precept—more forceful because acts are more impressive than words. Imitation is a protective process. A young animal which did not imitate its mother would have small chance of survival. When imitation is consciously stimulated by a feeling of respect it passes into obedience, and in this form powerfully contributes to our observance of discipline and morality, imposing a constant check upon our discretion in selecting methods of conduct.

Another force which influences our conduct unreflectively is that of *habit*. This is a chain of memorial associations which follow one another like camels in a caravan. We are conscious of them, but hardly realize their hold upon us, since it rests upon their familiarity, and of our acute susceptibility to this we are unaware. It is habit that takes a man to office of mornings, conducts him through the day, and brings him home again. It not only regulates most of our practical activities: it assists in safeguarding morality, since, when this has become conventional, no ideas[1] of conduct present them-

[1] 'Iam, non consilio bonus, sed more eo perductus, ut non tantum recte facere possim, sed nisi recte facere non possim.' (But one could not admire such a man.)

selves but such as are shaped by it. So by habitude a society becomes law-abiding and its citizens respectable. But virtue of this kind is a dead formality: it lacks the living resistance of psychic strength. 'The Letter killeth, but the Spirit giveth life.'

The guidance which we receive from imitation and habit leaves, however, a vast scope for expectative thought in the selection of methods of pursuit and avoidance. It may draw us from our habits. If, for instance, we habitually travel third-class, but, in thinking of an expected journey, recollect that it will fall upon a race day, we may decide to travel first-class to be less inconvenienced by the crowd. We think expectatively in regard to the multiform transactions of 'business'. For money-making depends in great measure upon probabilities, and these are taken into account when we reflect inferentially, and argue from sequences which vary in greater or less degree.

We turn now to *Explorative* thought. This is urged by curiosity, which, as we can feel, may be either involuntary or deliberate. It is clearly a resistant effort to the mental bewilderment that is caused by the new or strange, and, accordingly, it is analogous to courage.[3] The condition which it combats being mental, ideas must come into play. But when curiosity arises involuntarily or automatically, their action must be subconscious: they must affect our psychic energy unawares, by producing a state of bewilderment that evokes resistance. It will not be difficult to accept this conclusion if we realize how important is the part played by subconscious cerebration in the evolution of concepts, and even in the formation of inferential conclusions. By an evolutionary advance curiosity becomes voluntary when it is consciously actuated by ideas of the pride, or self-satisfaction, which will be won by its successful exercise.

The subjects which arouse our curiosity will depend upon our inquisitive susceptibilities. They vary greatly

[3] There is probably an affinity of origin between *cor* and *cura*.

from one person to another. Some are only interested in the affairs of domestic and social life; others extend their curiosity to politics, others again to science. But, as we have seen, *tastes* in curiosity may be cultivated by practice. Speaking generally, we are only interested in that which is connected with our existing knowledge, and are disposed to shut our eyes[4] to facts and arguments which open new lines of thought.

When curiosity is not confronted with a problem, it simply develops ideas by inferential unifications and associations: these come most easily in conversation, since there is then more than one spring to feed the current of thought. If in its course a strange conclusion is presented—one which is out of accord with our convictions—we resent its intrusion, and may very easily extend our dislike to him who introduced it. For it occasions an unpleasant doubt, which can only be dispelled by an effort of resistance. The art of pleasing conversation is, accordingly, to be picturesquely conventional. One who has a new doctrine to ventilate must disguise it by allying it with the imaginative charms of fiction, by investing it with the attraction of the mysterious, or by treating it humorously.

The problems which arise in explorative reflection are—Of what kind is it? What are its traits? What are the causes or consequences of its traits? and, What individual should be associated with certain traits? Thus we may be exercised concerning the nature of the criminal disposition, its leading features, its effect upon

[4] Thinking that it would humanize them, I sent on a visit to Calcutta the chiefs of a wild and troublesome hill-tribe inhabiting a remote part of the Assam frontier. They were conducted round the city and the port, and were shown factories, arsenals, and ships of war, but preserved a stolid silence. When I asked them what was the most impressive thing they had seen, they replied that it was the 'water running uphill', meaning the flood tide on the river. Their hill streams had familiarized them with the course of water. But they could not 'place' their unfamiliar experiences. This is not peculiar to the unlettered. There is truth in Anatole France's paradox that 'les savants ne sont pas curieux'.

the community, or in regard to the person who has taken some money off the dressing-table.

A desire to solve a problem stimulates a course of resistant or reasoning inference in which we endeavour to unify an observed trait with one of the elements of a sequence that has occurred in perceptive or reflective experience. So we may arrive at the kind or trait-class of an object, or at the happening which has preceded or will follow a happening. The process is of a judicative character, and the conclusion to which it leads is in fact a *judgement*. We may throw it into a syllogism of three thoughts—a reason of fact, a reason of experience, that is the sequence with which the fact is connected, and a conclusion by which the fact is incorporated in the sequence. The sequence is generally so familiar that, in explaining an inference, the reason of experience may be omitted. Syllogisms may be expressed in many verbal forms which Logic has endeavoured to tabulate completely. But its classification appeals so little to intelligence that, for centuries past, a *memoria technica*[5] has been employed to keep it in mind. Nor can it be said that the study of logic has contributed very materially to correct reasoning. For, in the first place, inference is in great measure subconscious, conclusions often presenting themselves before we realize their reasons. And, secondly, its course is continually deflected by emotional influences—by faith, prejudice, and imagination—against which logical precepts can hardly prevail.

We identify an object, respecting whose affinities we are curious, by noting a trait which is invariably possessed by a kind-class. If we know, for instance, that all ranunculous plants have acrid juice, and we meet with a plant having acrid juice, we may decide that it is of the ranunculous kind, since this trait is associated with the kind by an invariable sequence. So we *unify* the plant with its kind. That plants with acrid juice are ranunculi may appear to be a coincidence rather than

[5] The doggerel Latin lines beginning 'Barbara Celarent Darii'.

a sequence. But it presents itself in thought as a sequence, since thought is evolved from perception, and in perception the traits of an object or a scene present themselves successively,[6] not simultaneously. We may reason through a chain of such reflective sequences, as that feeling is internal sensation ; sensation is the appreciation of external vibrations ; *therefore*, feeling is the appreciation of internal vibrations, with the further consequence that internal vibrations must exist. Arguments from invariability of sequence may give a very high degree of probability. But the conclusion is not *certain* unless the sequence is peculiar as well as invariable. For the fact that *all* ranunculi possess acrid juice leaves it possible that this trait may be possessed by plants which are not ranunculi ; and the fact that *only* ranunculi possess acrid juice leaves it possible that there may be some ranunculi which do not possess it. But if the ranunculi *all* and *alone* possess acrid juice, any plant possessing it *must* be a ranunculus. Accordingly the correctness of our conclusions depends upon the breadth and exactitude of our knowledge—that is to say, of our experience. One who knows nothing of flowers may mistake a hollyhock for a rose, because it resembles a rose in form and colour. The Linnaean system of classifying plants was abandoned for the Natural system, as an increasing acquaintance with botany showed that the traits upon which it rested were often superficial.

We may assist our inquiry by the explorative activity of experiment, following upon a hypothesis. If, for example, we are in doubt as to the nature of a geological specimen, we may reflect that, if it is limestone, it will effervesce under hydrochloric acid, and may proceed to a practical test. Accordingly a hypothesis, followed by experiment, enables us to verify a conclusion. But in this case again we must be sure that limestone is

[6] So a sentence such as ' he was late again this morning for breakfast ' expresses a series of coincidences in the form of sequences. The leaves of a bud are in coincidence ; but they open in sequence.

peculiar in its reaction to the acid, as well as invariable.

When an object has been unified with a kind, or classified, it will possess the essential traits of its kind : if, for instance, we know that the plant is a *Solanum*, we know that some part of it is poisonous. If we proceed to inquire as to the causes or consequences of its trait, we rely upon a more obvious sequence. Traits, as we have seen, may take the form of changes or happenings. It is a relative trait of lightning that it is followed by thunder, and a sub-trait that the following is immediate. Accordingly, by means of consequential happenings, reasoning inference enables us to predict the future. We infer its course by connecting traits of the present with sequences that have been experienced. In everyday life we can predict the future in this fashion remarkably well, except in regard to the weather, the observed sequences of which are apparently vitiated by the omission of unperceived elements. But when dealing with the future of a nation—that is to say, with the consequences of its circumstances, its politics or its religion—sequences appear to fail us. We can argue correctly enough as to the future of animals and plants, since their life is one of instinct, or habit, and follows a regular order ; and, so far as the conduct of mankind is habitual, it is regular and can be predicted. But there is a disturbing fact in the immense influence of imagination, which sets up dazzling but unsubstantial ideas that may be quite out of accord with experience, and infuse an unpractical element into human motives. Their influence would, however, be small were it not spread from one mind to another by suggestion—that is to say through language. In this fashion the effects of such psychic stimuli as ideas of power—divine, clerical, regal, or numerical—of ascetic self-conquest, or of liberty, may be disseminated far and wide. A few men, or a single individual, who can command others, proselytize them, or persuade them can by revolutionizing their motives, bring about abnormal consequences. They may lead multitudes into courses which would never be suggested by reasoning from experience. Hence it comes that the lines of a nation's development cannot be predicted from its material surroundings. But we realize that there are laws: we say that 'history repeats itself', meaning, not only that similar circumstances are likely to produce similar results, but that the effects of suggestion follow a regular course and sooner or later fade away. The stability of political institutions, for instance, is dependent upon the existence of respect for them, and this is gradually weakened by disillusionment. Were these laws formulated, reasoning might steady the vacillations of political prophets. But serious difficulties would still remain. Reason acts through knowledge, and our information regarding both the past and the present is incomplete and inaccurate. Few histories are unbiassed; and the facts that are given us by newspapers are coloured, more or less, by propagandist motives. And reasoning in regard to national consequences can seldom be dispassionate. It is distracted by our political prejudices, by our pride, and by our hopes.

We also argue from sequences in determining the causes of particular happenings. We commonly draw conclusions from the regular sequences of habit. Thus we argue that there must be something wrong with a horse's bit or saddle, because he is restive, being ordinarily quiet, or that one who is not down to breakfast must be ill, because he is ordinarily punctual. A cause is something which always precedes happenings of a certain kind, and is invested by us with an idea of stimulating activity. It may be so invested simply because it precedes the happening: there are persons who think that bad weather is caused by a fall in the barometer. But as knowledge extends, an antecedent is not regarded as a cause unless its connexion with its consequent can be referred to a generalized sequence—that is to say, to a *law*. So we shift our idea of causality back from the barometer to changes of air-pressure, or of solar radiation, because these can be connected with changes of weather by physical laws. When, however, no regular connexion can be established in this fashion, we take refuge in 'chance' or 'luck', regarding it with an implication of causality, as when we say 'I missed my train *by* bad luck'. The primitive idea of a cause is simply that of an antecedent, as is shown by the use of the word 'because' to introduce a thought which leads up to a conclusion. Accordingly, before conceptions of universality and peculiarity are developed, or any knowledge of physical laws is acquired, a casual antecedent may be regarded as a cause, especially when the association of it with its consequence is enforced by an emotion of joy or sorrow. In this fact we discover the origin of unreasonable associations—of superstitions which may be merely fatuous, as that ill-luck follows the spilling of salt, or sailing on a Friday, but have also been the origin of prejudices which have blinded the intelligence of mankind.

The problem involved in connecting an unknown individual with a happening, is that which is commonly to be solved in criminal investigations. In this case the order of our reasons is reversed : we begin with the reason of experience—that is, with a generalized sequence, because we have before us no trait of character that can be attached to an individual. We have, however, a trait of relationship which introduces a generalized sequence into thought. If, for instance, I miss half-a-crown off the dressing-table, I may conclude that the housemaid must have taken it, because its taking *must have involved entering the room*, and she alone has entered it.

This is *deductive* as opposed to *inductive* reasoning : it opens with a reason (or sequence) of experience, and reaches its conclusion through a reason of fact—that the housemaid alone had entered the room. There being no trait of character to connect the incident with an individual, inference can run only by reversing its course; and, in exploring a fact, this appears to be the only case in which the deductive substitutes itself for the inductive process. But, when we are arguing from a hypothesis, as in mathematics, are generalizing the sequences of experience, as in philosophy, are demonstrating a conclusion to others, or are simply exercising our logical faculties, deductive reasoning may be used irrespective of fact: we can show that Socrates is mortal by commencing with the general sequence, or major premiss, that 'all men are mortal'. The deductive process is very largely employed as an instrument of persuasion, since it gains impressiveness by emphasizing generalities. In explaining a deductive conclusion, stress is laid upon the reason of experience, whereas an inductive conclusion is explained by its reason of fact. In itself, deductive is as reliable as inductive reasoning; either of them, if affected by faith, prejudice, or imagination, leads to a fallacy under the cover of logical form. But by emphasizing generalities the deductive process tends to put criticism off its guard: *dolus latet in generalibus*. If, for instance, one assures an audience that 'democracy is government in accordance with popular wishes, you belong to a democracy, and are therefore *free*'—(that is to say, are governed in accordance with your wishes)—he may induce his listeners to forget that, if they are out-voted, they have lost their freedom.

We also employ reasoning inference to ascertain the actuality of happenings which we have not witnessed. It uses the scales of experience. Our conclusions in these cases are in reality beliefs, since they rest upon the trust which we place in our informant. But we employ reasons of fact and of experience in testing his credibility. He may not be trustworthy, either because he was misled by his senses or by another person's suggestions, or because his experiences are misrepresented in words owing to prejudice or deceit. His statements are tested by comparison with our own knowledge of the consequences of human nature: if in the light of experiences they are probable, we do not draw largely upon human nature for arguments against their acceptance. If they are improbable, we do not believe them until we have subjected the witness's powers of observation, motives, and character to such criticisms as experience suggests to us. If, for instance, he is shown to be prejudiced, we associate his statements with the effects which in our own experience are produced by self-interest in the garbling of evidence.

Reasoning inference may be based upon sequences which are distorted by prejudice or are imagined, and in this case its conclusions will be prejudiced or fantastic. Conclusions will be true—or *trustworthy*—only when the sequences which lead to them have been derived from repeated experience. How far, it may be demanded, is experience itself a true indication of reality—of 'things as they are'? It can ultimately be resolved into conscious impressions of things within us or outside us—that is to say, into impression-ideas. These cannot afford us true conceptions of the vibratory conditions that are their origin, except upon the supposition that sensory nervous excitement is identical with the vibrations that arouse it. This appears to be impossible in itself, and can in some cases be proved to be incorrect. Science can infer, for instance, that light and sound are merely the effect upon the brain of vibrations that are dark and noiseless. Our ideas are phantasies. But they indirectly afford us some true knowledge of the real. It is certain, for example, that similarities and differences exist in our environment, that our ideas of changes and continuings arise from real changes and continuings, although they mislead us as to their character, and that, when changes occur in an invariable sequence, there must be between them some such relation as that of cause and consequence. The rhythmic succession of changes must be based upon reality, and so also must be the relationships that are observed between one impression and another. Mathematics can accordingly claim that they are concerned with realities; and Science in classifying impressions, and ascribing them to causes, is dealing with real similarities, differences and sequences, though of a very different nature to that which they present, as phenomena, when they are registered by the brain.

How, it may be asked, can the brain work in accordance with the requirements of correct reasoning unless it is guided by a higher intelligence? Through the control which is incessantly exercised by doubts. If, for instance, when a conclusion is based upon a peculiarity, a recollection comes forward which shows that the peculiarity is non-existent, the connexion is immediately reversed. Doubts are expressed in language as objections. They may invalidate a conclusion by showing that the reason of fact is erroneous, or that the reason of experience fails in not being invariable, universal, or peculiar. That is to say they suggest exceptions to the generalization upon which the conclusion is based. Objections may be answered by proof or demonstration. This is an effort of trial. How do we prove that a magnet attracts iron, or that 'righteousness exalteth a nation'? Simply by referring them to experience. In the case of the magnet we can verify our conclusions by the artificial experience of experiment. We cannot experiment with a nation, and consequently must use the experience which is recorded in history. We review the past to ascertain whether the righteous are always successful or respected.

Reasoning inference is, then, a process of the brain which, given an intelligent appreciation of samenesses, will yield correct results within the limits of its knowledge. But, for this, it is necessary that it should be dispassionate and unemotional—should be uncontrolled by the promptings of instinct or the self-assertion of egotism, and uninfluenced by faith, prejudice, or imagination. Love is blind to the lessons of experience: unthinking courage has no eyes for them: they are overruled by faith, distorted by prejudice, and eclipsed by imaginative hopes and fears.

Faith—respect for another's power and confidence in his protection—manifests itself as a feeling, but arises, as we have seen, from neural conditions on the physical plane and can, therefore, dominate us with overmastering power. We may appear to lose it. But it remains as a respectful confidence in ourselves. Faith stifles inference. Our thoughts are obedient and our conclusions take the form of beliefs. These may be altogether unreflective, the fruits of 'simple' faith. Or they may rest upon a reasoned appreciation of the power of our master, limiting it to certain branches of knowledge in which we conclude that he excels. But in either case it is incompatible with reasoning inference. Newton must have abandoned belief in the material intervention of Providence before he could have discovered by inference the effect of gravity upon the motion of the planets.

A vast number of our conclusions are in reality beliefs. In childhood we accept without question the thoughts that are suggested by our mothers: indeed the inquisitiveness of children, whether as to the composition of the moon, or the origin of babies, may be satisfied in the absurdest fashion. As we grow older, faith becomes a reasoned partiality for our own opinions, or for those of men whom we respect, and if these have been correctly framed, we are assisted, as by stepping-stones, in the pursuit of knowledge. But if our authorities have erred, having overshot their experience, or trusted to their feelings, their bequests are not stepping-stones but stumbling-blocks, which may retard indefinitely our progress towards the light. If we are in doubt as to the cause of a bad harvest, we may accept in trustfulness the conclusion that it was non-observance of the Sabbath, since the two have been connected in experience. Such beliefs are responsible for the extraordinary errors which have darkened man's judgement, and have introduced into history so much that is cruel and grotesque.[7]

Prejudice arises from emotional feelings—from a desire or aversion, an affection or dislike. It may not kill reasoning inference, but it disables it—' warping the judgement '—by excluding, or rejecting, thoughts which go against our inclination, so that the possible becomes the certain. If one whom we dislike is at all unceremonious in his manners, we are apt to impute to him deliberate rudeness. So a nation at war is unable to see anything to the adversary's credit : it eagerly accepts anything to his discredit, and cannot realize the lesson taught by history that the enemy of to-day may be the ally of to-morrow. By a similar prejudice a jury may condemn an accused because they dislike him. On the other hand, we are often blind to the faults of those whom we like : a military man rejects any imputation to the discredit of the army with a chivalrous indifference to evidence. No one can be a judge of his own quarrel. Nor can he criticize ideas of his own, until some time has elapsed since he felt the pride of conceiving them. We are similarly biassed towards the ideas of our ' set ', our class, or our nation. This is, indeed, the essence of patriotism, which, however much it may invigorate, blinds us to the truth.

The effect of imagination in distorting the process of reasoning will be noticed in the chapter following. It is at once the consequence and the cause of emotion. The brain, as we shall see, thinks imaginatively when it is affected by psychic excitement, and its imaginative thoughts react upon the psychic plane, and enhance emotional activity. In this condition it may create

[7] It is believed that between the years 1608 and 1680 no less than 70,000 persons were executed for witchcraft—many of them upon the direction of so eminent a judge as Sir Matthew Hale. (Carpenter's *Mesmerism, Spiritualism*, &c., p. 61.)

imaginary existences and occurrences and form imaginary sequences from them. It may substitute vague hopes or fears for reasoned expectations. Attractive ideals take the place of experience, obscuring by their brilliancy the beaten track which is marked by the success and failures of the past. Imagination may even distort the course of reasoning from experience. For reasoned methods of conduct, when deduced from fanciful samenesses, evolve into the fantastic practices of magic.

We have seen, then, that reasoning inference argues solely from experience. In our practical life it endeavours to supersede instinctive and emotional influences but cannot exclude them. In our life of inquiry it maintains itself by a constant struggle against the rivalry of faith, prejudice, and imagination.

CHAPTER IX

APPRECIATIVE AND IMAGINATIVE THOUGHT

THE phases of thought which we have distinguished as *appreciative* and *imaginative* are initiated, not by purposeful impulses, but by conditions of pleasurable and displeasurable excitement which at first sight may appear to be merely passive. If, however, we examine the effect upon us of a pleasing or displeasing stimulus or idea, we shall find that it is infused with a certain activity—that of 'savouring' or appreciating the pleasure or displeasure. We term this feeling [1] 'like' or 'dislike', 'admiration' or 'contempt', according as the neural state which it represents arises on the physical plane and is reflected on the psychic, or is aroused on the physical plane through the psychic. Likes and dislikes are neural conditions of attraction or repulsion, such as in instinctive life excite approach and recoil: they are physical impulses which vent themselves upon the psychic plane. They are passive in that they do not lead to practical action: they are active in that they incline us [2] towards or away from a stimulus. When pleasure or displeasure is caused by a *psychic* stimulus, which arouses physical attraction or repulsion through psychic excitement, like becomes admiration, dislike contempt. But, so distinct are our physical and psychic natures, that one may like what he despises—an incongruity which distracts many unfortunate love

[1] This perception of activity is of comparatively recent date: 'like' in its original use expressed not our own activity but that of the stimulus: 'it likes me' as 'il me plaît'.

[2] 'All sensation, such as that of a rather lively external impression, is followed by a tendency to movement, if not by actual movement.' (Morat's *Physiology of the Nervous System*, p. 518.)

affairs. The psychic excitement that accompanies these appreciations may pass into *imaginative* activity—perhaps the most complicated of all our faculties. The imagination is also excited by the pleasure or displeasure which attends upon strong emotion, such as that of fear or love, or upon the psychic revulsions of joy and sorrow, pride and shame.

First of *Appreciative* thought. This is initiated by the pleasure or displeasure which is caused by a sensory impression, a feeling, a recollection, or an idea. We are attracted by the pleasurable and like it: repelled by the displeasurable and dislike it. Our like, or dislike, stimulates thought. We appreciate the attractive or repellent as the good or bad, the beautiful or ugly, the right or wrong, the true or false, according as it affects us physically, psychically, or reflectively. The good is favourable, the bad unfavourable to our physical life: a thing is beautiful or ugly, right or wrong, according as it conforms or does not conform to our standards of aesthetic taste or morality. These are of psychic origin since they are based upon ideas of excellence or of resisting power. The true accords with experience as utilized by reasoning: the false is in disaccord with it. Reason also enters into our appreciation of the good and bad, the beautiful or ugly, the right or wrong, since its recognition as such involves its classification. Hence if judgement is misled by prejudice or faith, we may confuse good with bad, right with wrong, truth with falsity. So, for instance, if we have an affection for another we may find something to like in his misdeeds or oppressiveness. Orientals admire a ruler who is strong and capricious: they style this the 'kingly disposition'. We condemn another's 'militarism', but not our own.

Purely physical likes and dislikes are few in number and of a simple nature. Such are our likes for certain touches, as caresses, for food, and for light: our dislikes of pain, hunger, and fatigue. They become infused with a psychic element when their stimuli are associated with ideas of excellence or inferiority, since these, being the antecedents of pride or shame, arouse psychic excitement. Like then becomes the appreciation of admiration (respectful, if excellence is *powerful*), dislike that of contempt. We shall find that it is ideas of excellence that give pleasure to our acquired, or differentiated, tastes, which are infinitely more numerous and complicated than purely physical pleasures. We term them 'aesthetic', appreciating that they are influenced by an element other than the physical. Beauty is excellence, ugliness inferiority, and since ideas of excellence and inferiority depend upon standards of excellence and these are variable, some find beauty in what to others is ugliness. It is almost impossible for a European to admire Japanese music: indeed there is much in modern music which would have been repellent to the taste of a century ago.

Admiration and contempt may be aroused by feelings as well as by external impressions. We have a great admiration for our own power and dignity. Courage is power and is admirable: cowardice is weakness and is despicable. Dexterity in action or speech is admired because it manifests power. We admire liberty because it is infused with the strength of self-confidence. There is the strength of resistance in purity and morality, the weakness of yielding in pollution and sin. Beauty may, of course, be moral as well as physical. We admire the 'beauty of holiness'.

It is important to realize that the appreciation of a pleasure or displeasure includes the appraisement, or valuation, of its *intensity*. Psychic excitement, whether expansive or contractive, varies greatly in degree, and hence pleasures and displeasures vary in poignancy. We appreciate these differences and remember them. It is by such appreciative recollections that the course of deliberate choice is guided: failing them, selective volition would be impossible.

The stimulus which is nearest us is oneself, and accordingly appreciative thought is very largely introspective or self-conscious. This, as we shall see, involves imagination, for one becomes self-conscious by personifying his psychic and physical traits as individualities that are distinct from his reflective self, so that they can excite like or dislike, admiration or contempt. We may honestly dislike or despise ourselves. But there is no one of sane mind who does not regard himself favourably on the whole. If, indeed, we were not endowed with self-assurance, we should fare badly in the struggle for life. Under the stress of psychic revulsion, generally brought about by the words of another, we may view our past conduct with the disgust of repentance. But we find something attractive in the very strength of our self-abasement. Our leaders are 'at one' with us as our representatives, and politics—that is to say, the conduct of our leaders and their opponents—provide inexhaustible material for appreciative thought.

Admiration and contempt involve strong psychic nervous excitement, and they introduce us to the consideration of *Imaginative* thought. We enter upon a most difficult subject. For the imaginative faculty is mysteriously pervasive in its influence: it affects behaviour as well as thought: it is the inspiration of Art: it ennobles Politics or degrades them. From the scientific point of view it possesses the extraordinary characteristic of running counter to the tendencies which are pressed upon us by evolution. We must not, then, search for its origin in any nervous activities whose consequences are materially, or practically, useful in the struggle for life.

We have repeatedly referred to our 'expressions of emotion', and may seem to have attached to them more importance than they deserve. But they are, indeed, very remarkable phases of muscular activity. They represent the 'play' of nervous excitement upon the body: emotions are converted into muscular movements, as when a musician expresses his feelings upon the piano. We recognize in common speech that their function is to relieve feelings: if they are repressed, nervous excitement increases, and may attain a violence that may shatter the strings of consciousness and even endanger life. They appear to have no other intrinsic utility. It may be urged that they are of value in attracting the sympathy or help of friends, or in frightening or 'warning-off' foes. But what can be the practical value of laughter and tears? Certain of our emotional expressions—utterances, in particular—have become of inestimable utility as instruments of social telegraphy, and owe their complicated development to the stimulus of this advantage. But it is not necessary to this end that they should be inherently purposeful: evolution abounds with instances of the utilization of incidental or by-products. If they are a purposeful outgrowth of social life, why do they affect us as strongly in solitude as in society? Because, it may be concluded, their primitive function is to act as safety-valves for emotional excitement. They are not, of course, peculiar to man. But in him they are far more complex than in the lower animals owing to the plasticity of the muscles of his throat, tongue and lips, and of his arms and fingers.

The effect of these manifestations is to convert a feeling into something tangible, for, as we have seen, muscular movements are felt as touches. It is a significant coincidence that, when the brain thinks under the influence of emotion, its thoughts become imaginative by assuming *tangible* forms. When Isaiah exclaimed that 'All flesh is grass' he gave tangible, or concrete, expression to the thought that all men are mortal. Imagined ideas are picturesque: they counterfeit percepts. As illustrated by the prophet our common mortality could be figured pictorially. As a general proposition, it could not be portrayed. Imagination, then, appears to be a mental phase of emotional expression—to be emotional thought evolved from the muscular expression of emotion, and shaping its ideas in accord with the nature of this muscular expression—and we shall discover further testimony to its evolution from expression when we examine its phases in more detail. Strong emotion is not, however, adequately relieved by the elaboration of concrete ideas in thought. It presses for their materialization in actually tangible form—in gestures or manners, in utterances or words, or in objects which serve as emblems of imagined ideas.

When our imagination is active we can perceive that we are affected emotionally. The use of imagery has been formalized by language and has become an art, which may be used in cold blood by an effort of assertive or tentative volition. But we are concerned with its evolution, and must limit our regard to its spontaneous manifestations. These are always associated with conditions of psychic expansion or contraction. The excitement may be aroused by the agreeableness or disagreeableness of a feature or trait which is presented by a percept or an idea ; and the more acute is the sensibility the more active is the imagining—a truth which is illustrated by the artistic temperament. But it may also be a mood which possesses us. The psychic excitement of love or fear conjures up fancies which restimulate these emotions. The hope that springs from enthusiasm begets pleasing images, and is assured by them. Pride and shame are emotional geysers which send up streams of exhilarating or depressing visions.

Imagination affects us in perception, in reflection, in reasoning and in action, and it is difficult to describe its phases simply and clearly. It materializes a trait that excites our admiration by *vivifying* or *personifying* it—that is to say, by endowing it with a living existence. This testifies to its evolutionary origin in emotional expression : for expressions are manifestations of *life*. If the trait belongs to an impression that is received in the course of perception, the effect of this process is to *idealize* perception : thus, if struck by the beauty of a sunrise, we think of 'rosy-fingered dawn',[3] we have idealized what is really a natural incident. If the trait occurs to us in thought, it is, on the contrary, *materialized* —that is to say, invested with a concrete form—by being personified: so we may figure Time as the 'thief' of good intentions. In both these cases a living individual is created, and we speak of imagination as 'creative' when it personifies. But imagination may also be decorative. If an object or idea excites our admiration we variegate or embellish it with ornamentation—that is to say, give a tangible form to its excellencies, just as we embellish ourselves with clothes and jewellery. So, if a winter landscape pleases us, we may imagine it as adorned with summer foliage. Here again imagination testifies to its origin: for expressions of emotion *adorn* the features with a charm of variety. Accordingly we may distinguish imaginings as creative and decorative, each of these phases being again subdivided according as it is perceptive or reflective.

Perceptive imagination in its most exaggerated form appears to be illustrated by dreaming. Conscious sensation is then suspended, but impressions, internal and external, may touch the physical plane unconsciously and may reach the brain circuitously, through the spinal cord. Dream images are all concrete—are, in fact, individualized ideas of unconscious nervous conditions which may be the relics of waking excitement, but appear to be generally caused by external or internal impressions that are received during sleep. It is well known that dreaming may be initiated by touches or sounds. But internal impressions, often of sexual[4] origin, are still more exciting. Dreams present the effects of these stimuli in the form of sensory impressions which possess a trait in common with them: thus in a con-

[3] It is 'by metaphor' that the dawn is rosy-fingered. But the metaphor is imaginative when it invests an inanimate object with a trait of life by endowing it with an ideal personality.

[4] An interesting point insisted upon by Freud, and over-emphasized by him.

dition of giddiness one would dream of the edge of a precipice.[5] If an incident has shaken our nervous equilibrium, its residual effects will become dreams by arousing psychic excitement that stimulates the brain imaginatively as well as memorially. The strangeness of our dream-fancies is partly due to their recall of record-ideas which have been acquired subconsciously. But the chief cause of their erratic extravagances is that their stimuli initiate unificatings and associatings which, being unchecked by differences, may carry us very far indeed from the impression, or nervous condition, which originated the mental disturbance. A disregard of actual differences is a characteristic of imagination. The dawn is not a rosy-fingered damsel, time is not a thief; but the imagining brain seizes hold of a sameness and uses it to unify a trait or an object with an idea which, from the point of view of sensation, is essentially dissimilar. The excitement which urges it appears to prevent the reversals that are ordinarily caused by differences. Accordingly, imagination obscures sensation, and when sensation is feeble, it may run riot. Such may be our experiences during the dark hours of a sleepless night. We are haunted by ghosts, which fly away at cock crow, when daylight brings conscious sensation into the fullness of its activity, and we see things 'in their true light'.

Creative imagination is illustrated by self-consciousness. In this condition one is not merely conscious of himself: he figures other personalities within himself, by personifying the traits of his physical and psychic natures, and can actually imagine that one is in personal conflict with the other. This occurs during 'tempta-

[5] Here is a curious illustration from my own experience. Making a night railway journey through Egypt, I could only find room to lie down by putting my feet out of the window. I dreamt that while bathing in the Nile, my foot was seized by a crocodile, and awoke to find that a friend in the next compartment was amusing himself by trying to pull one of my shoes off. This produced an unconscious impression of gripping, which the crocodile illustrated

tion ', when psychic resist physical propensities. If the temptation is overcome, the self-conscious man enjoys an indirect pride, apart from that which is the conscious presentment of the expansion that follows successful effort. For the succeeding reaction is figured by him as the respect of his physical personality, and affords the pride that is linked in reversed sequence with respect. The curious aberration of ' multiple personality ' simply exaggerates a subdivision of self, which is an ordinary feature of imaginative activity.

Imagination reaches its climax in hallucination. We have seen that nervous associations in sequence may be reversed. Hence fancies may produce such conditions of the sensory organs as have accompanied similar impressions in the past: sensation is associatively stimulated *backwards* by the brain. Hallucinations are particularly vivid in the nervous conditions of trance, fascination, or hypnotic submission, when the brain is overcome by a flush of psychic excitement. This may be caused by an over-exciting idea. If it is presented in sensation it may produce fascination: if it is presented in thought, it may be followed by a succession of images, so vivid as to be mistaken for sensory impressions. When imaginative activity is very acute it may pass into insanity —a condition of enduring hallucination—and we may suspect that mental derangement is connected with an abnormal development of cerebellar influence upon the brain.

We have been illustrating some abnormal phases of creative imagination. But its normal activities follow similar lines. They involve the particularization of a trait by personifying it. In regarding a birch tree overhanging a torrent we may think that it is like a fairy because it is graceful. Imagination goes farther, and sees it as the ' fairy of the glen '. So Shelley figures the West Wind as an enchanter—because the dead leaves, like ghosts, flee before it. The sameness between the wind and an enchanter is only in a single trait. But this is the feature which moves us, and the imagination unifies it with an individual that possesses it, in disregard of actual differences. It acts upon analogies that are partial and incomplete.

Accordingly imagination may give life to the lifeless. To a little child its favourite toys are alive,[6] because their traits are identified with life. The savage attributes a living individuality to natural objects which arouse emotion by their strangeness or fearfulness. This is Animism, man's first step towards the temple of the Unseen. By imaginative inferences, such as that which is based upon the similarity between man's breath and the wind, simple animistic beliefs develop into more elaborate creeds. We shall touch upon this point again before closing the chapter.

To creative imagination that is stimulated by reflection we can trace the origin of our Ideals and of Art. A guiding ideal is obviously the personified abstract of a stimulating psychic trait. We think of it as a personality.[7] Purity, Morality, Liberty, Cruelty, and Hate are figured as actual existences—as divinities to whom we can render service. They are inspiring because they represent power in one or other of its implications; and accordingly give dignity to life if associated with ourselves. And, since they may run counter to physical promptings—and often to the lessons of experience—they appear to raise human conduct above its natural evolutionary plane. We *pursue* them by expressing them in our behaviour. High ideals attract our admiration. But they may misguide us sadly. Nevertheless they have been of incalculable service to man in leading him to venture upon the unknown. If they are not endorsed by reason—that is to say, if experience shows that they offer no real advantage—they are abandoned: man's history abounds in ideals that have

[6] Strikingly illustrated by the little boy who was found showing a picture-book to his rocking-horse.

[7] And, accordingly, give it the distinction of a capital letter.

sprung and faded. In these days we cannot understand the enthusiasm of the Crusades; and asceticism, as a good in itself, is losing its past glories. But they may be approved by reason as actually beneficial: in this case they are a progressive evolutionary force. We owe to them the development of morality from the magical notions of primitive superstition.

Artistic ideals are personifications of aesthetic traits —that is to say of kinds of excellencies. Beauty is personified by the idea of it which the sculptor expresses in a statue. A group of traits may present themselves together in thought and be personified as a whole. Strength and swiftness may be figured as a winged lion. The characters of the drama and fiction are such composite personalities, each animating a complex of traits of character, drawn from experience, and heightened imaginatively. They are 'dramatized' by a succession of actions, speeches, and incidents of which they are figured as the causes or consequences. These are traits of relationship—drawn from actual life and fancifully embellished—which are vivified by being attached to the personalities. Imagination does not exclude inference, and Art uses experience as the foundation upon which it builds its fantasies. Its characters are *made*, but its incidents are mainly derived from memorial associations—indeed, its heroes and heroines may be designed from recollections of actual individuals. It may even accept experience as its *motif*, and set itself to illustrate imaginatively truths concerning the course of human passions which are idealized as 'laws' of human nature. Fiction of this class is in reality imaginative science: it is an expansion of the parable.

Characters that are imaginatively personified are primarily ideas of *kind.* But they are not felt to be complete until they are particularized by proper names.[8]

[8] Every one who has tried to amuse children by story-telling knows how pertinaciously they insist upon the *naming* of the characters. So particularized they become tangible.

Here again we can trace the footprints of evolution. For expressions of emotion, although common in kind to all humanity, have in them that which distinguishes the individual.

We have been assuming that artists and authors are 'enthused' by the ideas which they depict, and it may be objected that it is possible for one to compose in cold blood. He may *imitate*—may make excerpts from experience, or from other artistic compositions, and combine them with dexterity. But he will not *create* except when touched by the glow of the 'sacred fire'. It is, however, to be observed that all Art is in some degree imitative: the artist must draw from his surroundings the filigree of relationships in which he sets his characters—their dress, manners, language, and adventures. Accordingly Art progresses by evolution, and its development can be studied historically.

Let us pass now more particularly to imaginings which embellish or decorate. We subconsciously add to the attractions of an individual or object which pleases us, and to the repulsiveness of one that displeases us, by materializing our feelings in imaginary traits. To a lover the charms of his mistress vastly surpass reality: love is, indeed, fed by imagination, since this adds excellencies which, although fanciful, stimulate an accession of admiration. On the other hand an enemy is invested with vices which are drawn from fancy. We illustrate this process in ourselves. If one is in low spirits he may readily associate the idea of sickness with himself, and 'fancy himself ill'—or think of himself as abandoned by his friends and alone in the world—associations which may move him to tears of self-pity. On the other hand, in good spirits, one may easily figure himself as winning money or lowering his golf handicap.

The artist similarly decorates the creatures of his fancy, and, as it is said, 'works up the details' of his conceptions. In words one glorifies an idea by praising it: in deeds by beautifying its emblem or symbol.

Decorative ideas are suggested to the artist by his likes : they are recollections which are associated with likings, and may be classed as shapes, colours, and rhythms, according as they have impressed themselves on him through touch, sight, or hearing. An ornamental pattern is always rhythmic, and when rhythm is applied to verbal expression it invests prose with the form of poetry. Decoration expresses these embellishing ideas by representing one or other of their features. A decorative acanthus leaf is not a leaf, but the shape of the leaf ; and, since this is ideal, it may be varied by imaginative evolution, or be degraded by unskilful imitation. Decoration may be applied to manners, transforming them into fantastic ceremonies.[9] It has endowed language with literary grace, and rendered it a polished instrument of expression.

We saw at the beginning of this chapter that emotion may not liberate itself completely by the creation of tangible images in the brain : it may press for the realization of these ideas in things which are actually tangible. The artist is 'forced to express himself' in action or speech. The formation of mental images can, then, hardly be considered apart from their expression, and in the preceding pages it has been impossible to exclude all reference to imaginative action. This is a subject to be treated in detail when we come to consider conduct and behaviour. But we must briefly refer to it here.

Imaginative fantasies insist upon being expressed when the pleasure or displeasure from which they arise overbalance the practical aims of life. This is so generally during childhood. And keen susceptibility to the pleasing and displeasing persists throughout life with those who possess the artistic temperament. Expression involves the materialization of an idea as an impression of touch ; and, since acts and words involve touches, images may materialize themselves in manners,

[9] Ceremonies are termed 'fantasias' in modern Greek and Arabic.

speech, and writing, as well as through the fashioning of objects. They are expressed, for example, by acting as well as by painting. Indeed the representation of an imagined personality by dramatic action appears to be the most primitive means of imaginative expression: we can observe it in the play of puppies and kittens. Expressions of imagined ideas are always emblematic, for they represent a whole by a part of it. An individual is represented by a trait. A statue or picture is not the thing which it emblematizes, but an embodiment of certain features of the thing. It is, then, to be noted that the expression of an image follows a course that is the contrary of that through which it is conceived.

An imagined idea may be tangibly materialized by the adoption of an object as well as by the making of it. This is illustrated in the play of children—the simplest form of imaginative expression. To children 'playing at bears' the sofa is an emblem of a cave, while their gestures are emblems of the bear and its hunters. An imaginative emblem should have some resemblance, however fanciful, to the idea for which it stands; and, speaking generally, the closer the similarity, the better does it fulfil its purpose. The *making* of an emblem is a long step in advance: it involves manipulation which imitates the idea by outlining its shape, as when a painter foreshadows his sketch by moving his pencil in the air before him. Such imitative gestures [10] applied to a surface that retained a record of them would be the commencement of drawing. These developments, however, are for consideration at a later stage. But we may note that an emblem or symbol that expresses the idea of one person is an imaginative stimulus to another. Thus a picture or keepsake is at once the consequence and the cause of imaginative activity.

[10] They are largely employed by deaf-mutes in communicating with one another.

We have seen that imagining and reasoning may progress together—that reasoning may be imaginative. We may reason from reflective sequences that are derived from fancied existences, traits, and occurrences: of this kind were the disputations which exercised intelligence during the Middle Ages. Coloured by likes and dislikes reasoned expectations become hopes and fears, based, not upon experience, but upon images that 'parabolize' nervous conditions. If we are in good spirits we regard the future hopefully; if in bad spirits, despairingly. And imagination may mislead the process of reasoning when it is concerned with the methods by which we may realize our hopes, or escape from our fears. It lures the brain into unifying ideas through fanciful samenesses. This is illustrated very strikingly by the practice of magic. It is the product of *expectative* thought, since it is an instrument of pursuit or avoidance. But it rests upon fantastic identities. There is such an identity between a figure representing a man and the man, and accordingly it is imagined that one can afflict another by mutilating a likeness of him. In one of the Indian hill tribes mothers who wish their children to grow fat roll them in a pigsty. A Hindu who has excavated a tank formally 'marries' it to a stake that is set up in its midst—a similar ceremony may be performed for a newly planted grove—the inference being that as marriage is followed by children it will render the tank productive of fish and the orchard of fruit. A natural object is representative of its *kind*: a tribe is a kind, and the object may accordingly be adopted as a tribal *totem*. One fanciful identity slides easily into another, and magic rites may lose all apparent connexion with the analogy out of which they sprung. But they are plainly the fruits of imaginative reasoning. They carry us back to days which seem remote in the evolutionary history of civilized man. But they survive, unsuspectedly, in many modern observances.

The effect of imagination upon *explorative* thought has been more momentous and more enduring. There is, as has been noted, an identity between the breath and the wind, and since this does not die, the spirit of man may be immortal. Hence the dead may live in spirit—an idea which immediately leads to a belief in ghosts—that is to say, in the Unseen. Moreover, if man's life can be dissociated from his body, it follows that life itself may be spiritual, that all things possess souls apart from their bodies. Ideas of their spirits develop into those of particular personalities, such as gnomes and dryads, gods and goddesses. If the feelings that are the traits of our own personality are personified in this fashion, there come into existence the 'emotional divinities'[11] that characterized the classical Pantheon. In Mars courage was deified, in Venus love, in Minerva wisdom. And since we cannot conceive of a happening without a cause, there must be a First Cause of the inexplicable wonders of Nature, that is personified as the Creator and Governor of the Universe. Being regarded anthropomorphically, he will be pleased or displeased by what pleases and displeases ourselves—by the moral resistances and renunciations which afford pride and the lapses which are attended by shame, and will be gratified by such behaviour as would gratify ourselves. Accordingly his service may include morality as well as propitiatory adulation and offerings: it may be embellished by ceremonies and committed to a hierarchy.

We have reviewed in this chapter the effect upon thought of nervous conditions of pleasure and displeasure. If they touch the physical plane of our nature they become likes and dislikes, which pass into the stronger feelings of admiration and contempt when the psychic plane is also excited, as is the case if the pleasing or displeasing stimuli owe their character to

[11] The older 'Titanic' divinities, that were overthrown by them, represented forces of Nature.

an implication of superiority or inferiority. Likes and dislikes affect our judgement as prejudices : they transform reasoned expectations into hopes and fears. Psychic excitement, pleasurable or displeasurable, has a momentous consequence : it arouses imaginative activity—an evolution from the *consequences* of psychic excitement in emotional expression—which may affect perception, action, and speech as well as thought. Imagination converts ideas of pleasing and displeasing traits into ideas of tangible individuals, and embellishes ideas of individuals with tangible decorations. So it creates personalities within ourselves and personifies stimulating traits as Ideals. By enhancing the effect of similarities and analogies and obscuring that of differences it profoundly modifies the course of reasoning inference. It expresses itself in Play, and is the ultimate source of Art and Religion.

CHAPTER X

THOUGHT AND MUSCULAR MOVEMENT

THE young of brainless animals are, for the most part, born in a state of competent activity: they come into the world 'ready-made' to meet its difficulties. Of the higher animals it may be said that the better developed is their brain, the more incapable is their infancy. Young fishes and reptiles can shift for themselves immediately they are hatched. Amongst the birds the mother must become a nurse for the education of her young, and the period of nursery life is lengthened amongst the quadrupeds. We can, therefore, infer that instinctive, or reflex, capacity is weakened by the evolution of a brain, and that a brain requires a period of infancy during which it can be stocked with ideas. For the behaviour of birds and beasts is plainly guided by ideas. We term it 'instinctive'. But it is not *reflex*. We know that with ourselves an action is preceded by an idea of it. A notion of what we are going to do or say occurs to us before we do or say it. When we speak of 'doing a thing', 'making a thing', or 'saying a thing' the word *thing*[1] signifies an idea that is before us. It may be a very general idea, for the particular movements involved in an action or utterance, once learnt, are repeated memorially. But an infant learns its first steps by carefully—though, it may be, subconsciously—forming ideas of its tentative movements. Accordingly, it is through *ideas* of acts and utterances that conduct and expression are brought under the control of the brain. The ideas may materialize themselves automatically through association: or, guided by them,

[1] 'Thing' is derived from A.S. *thenken*, to appear, used in the expression 'me-thinks'. An idea is an *appearance*.

an effort of will may substitute itself for the physical or psychic neural states of which acts and utterances are primitively the instinctive consequences. Instead of being impelled to cough by repulsive nervous excitement, we can cough deliberately through an *idea of coughing.*

It is clear that ideas of movements can only be formed from actual movements, and it follows that the earliest of our actions and utterances must have been idealess and unconscious. They may be classed as purposeful reflexes, as automatic expressions of feeling, or as efforts of venture. The first are *practical,* being directed towards an object; the second are *manifestative,* simply expressing feeling; the third may be of one class or the other. As soon as ideas are formed of them, they may be elaborated and controlled by the memorial and intelligent faculties of the brain: efforts of venture, in particular, are guided and developed by familiarities of trait between expectation and movement, and by recollections of their consequences. Once formed, ideas of movements are associated, isolated, and generalized, like other ideas, and give rise to concepts of actions, manners, and courses of conduct. As growth advances, these concepts are executed in detail by the unconscious association of motor nerve-cells—that is to say, muscular activity arises in unconsciousness, and passes through consciousness to unconsciousness again.

Practical reflexes are represented in man by such instinctive movements as the grippings, suckings, and swallowings of infancy, and involuntary movements of approach and recoil. They are but feebly developed, and furnish few ideas to him. While they supply the *motives* of his appetites, of his strongest emotions, his likes and dislikes, pursuits and avoidances, they suggest no particular means of obtaining satisfaction, and, being stereotyped by instinct, are hardly capable of being elaborated by individual evolution.

The most primitive manifestative movements are the smiles and frowns, gestures and utterances which relieve and express the feelings of the moment. But the simplest of our manners also belong to this class. A feeling of pride automatically manifests itself in stiffness of deportment—or a strut; humility by a relaxation of the body muscles. All these expressions are primitively unconscious. But ideas are formed by them which are capable of great development, since manifestative movements being purposeless, are plastic. So have arisen, for instance, the very complicated mannerisms by which pride and humility may be manifested.

Efforts of venture, which seem to promise so little, have, however, been the most fruitful material for evolution, because they are wholly undirected by instinct and are therefore 'free'. Primitively of a random character, they leave imprinted on the brain an idea of the movement made and of its consequences. These may often be accidentally useful; and when we come to discuss the evolution of methods of conduct, we shall see how greatly it is indebted to fortuitous sequences. If the consequences include a trait of relationship, or analogy, which unifies them with an expectation or desire, the movement can be repeated so as to increase this sameness. If, for instance, a jump falls short, its length can be increased by the introduction of an idea of greater intensity of effort, or of a running start. Efforts of trial can, therefore, be directed and improved by intelligence: they can be stimulated by ideas of movement which have been amplified by inference—in fact, they become thoughtful instead of thoughtless—and each successful trial stamps itself upon the brain as an idea of useful movement in certain relationships. Trial becomes practice when it is repeated in order to assimilate action with inferred expectation, and practice is a potent instrument for the attainment of skill. Efforts are constantly being stimulated by difficulties: 'necessity is the mother of invention'. In order to climb trees man had no need of a climbing reflex, for ideas of climbing would be evolved from endeavours to reach an attractive fruit, guided by a sameness of relationship between expectation and movement. The fruit being *above*, the movement must be *upwards*.

It appears, then, that man owes his complexity of behaviour partly to his freedom from instinctive control, and partly to the intelligence that enables him to appreciate samenesses of relationship, and to generalize ideas. Beasts and birds form ideas of actions. Their behaviour is not instinctive in the sense in which this term applies to brainless creatures. But their ideas are mainly of reflexes, and can hardly depart from instinctive lines. They may, it is true, acquire new ideas of conduct by efforts of venture: a bird may change its nesting habits, and a dog learn to sit up. But these novelties are forced and cannot endure. And the lower animals appear to have little faculty for generalization. A dog which has learnt to open a door latch cannot apply its knowledge to a situation which varies from that in which it gained its experience. It cannot think of its action as isolated from its particular relationships or circumstances, and cannot, therefore, practise it. Man, on the other hand, sedulously practises new movements and thereby perfects them.

An idea of a movement is followed by the movement. We may wonder how one can stimulate, or cause, the other. For the function of the motor nerve-cells is to respond to *actual* stimuli.[2] How can they be set in action by ideas of the brain? The intimate association of nerve-cells by familiarity of rhythm will furnish us with a clue to the mystery. We have seen that an idea of an injury will evoke a wince because, having

[2] They shorten the length of a muscle so that it exerts a *pull*. We *push* by pulling out the limb between the object and ourselves, or between it and the ground. The shortening results from a change in the shape of the muscle-cells: they expand from side to side, and, since they do not alter in bulk, this involves their flattening from back to front. A shortening would also result did the muscle-cells *contract* from back to front. Hence two contrary phases of motor excitement would produce similar movements.

been associated with an injury, it is also associated in sequence with the muscular consequences of an injury. Ideas of movement appear to stimulate movement in a similar fashion. But the sequence is reversed. They stimulate causes instead of consequences. An idea of a step has been associated coincidently with a step: it is, therefore, associated in reversed sequence with the action of the motor nerve-cells which actuated the step. The reversal of a nervous sequence has been repeatedly illustrated; but it is a process which defeats perceptive analysis, and can only be detected by inference.

We may appear to be attributing too much to nerve association. But our conclusion is fortified by the curious facts of auto-suggestion, in which one physical or nervous condition produces another with which it has been associated in rhythm. Kindly acts are associated in experience with kindly feelings, smiles with pleasure, frowns with displeasure. Accordingly, if we act in kindly fashion, although with no sympathetic motive, a glow of kindly feeling may follow[3]; and by deliberately smiling or frowning we can conjure up faint shadows of pleasure and displeasure. There can be no doubt that the conventional politenesses, upon which society insists, go some way to promote actual good feeling.

Auto-suggestion culminates in hysteria, the pathology of which abounds with the strangest illustrations of the power of an obsessing idea to produce the physical conditions from which it, or one of its traits, has been formed. The sufferers may not only produce derangements of the kind ordinarily termed 'nervous': they may inflict actual injuries upon themselves—may force the development of tumours[4] and of the remarkable ex-

[3] These associative interactions appear to have contributed largely to the foundations of the Pragmatic school of Psychology.

[4] An amusing illustration is given in the *Lancet* of June 1880. A gentleman thought that he had swallowed his false teeth. He felt them in his throat, and a hard swelling appeared behind the larynx. A surgeon was summoned, and was about to operate, when a servant rushed in with the teeth in her hand. She had found them in a drawer. The swelling subsided immediately.

ternal wounds known as 'stigmata'. It is well known that men can think themselves into simulated cholera and lesser abdominal troubles. And, on the other hand, a lively idea of recovery, especially if based upon faith, will actually materialize itself in recovery—a fact which is demonstrated by the triumphs of psychiatry,[6] and the indubitable successes of Christian Science.

Ideas of acts or utterances may pass into muscular movement by an *effort*—it may be of restless or petulant impatience that opposes itself to, or contrarifies, suggestions that are offered by the brain, or by the words of others—it may be an effort of trial in resistance to doubt. These are phases of 'willing', the consideration of which may conveniently be postponed to the chapter following. Apart from them, ideas of movement, as we have seen, materialize themselves *automatically*, through association, unless expression is resisted or inhibited by volition because its consequences would be inconvenient. A convincing proof of this is the habit of talking to oneself or thinking aloud. Each idea of utterance, as it presents itself, is immediately expressed. Children are restless, troublesome, and garrulous because their ideas are instantly materialized. A schoolboy's careless whistling automatically expresses musical ideas that float through his brain. In somnambulism subconscious ideas of action pass directly into action. The automatic expression of ideas is also illustrated by subconscious imitation—a process which has already been touched upon. Yielding to imitative expression we may join in meaningless cheering, and in applauding performances which we do not really appreciate. Impresarios have learnt that manifestations may be drawn from an audience by the example of professional claqueurs.

When actions or utterances have been practised a

[6] In the Neurological Hospital at Newton Abbot, 96 of 100 soldiers, who had been completely incapacitated for an average period of eleven months, were cured in an average of fifty-four minutes (*Nature*, February 18, 1919).

notable development occurs. The multitudinous movements that have been learnt by experiment are associated in varied successions, as they have been combined in practice, and can repeat themselves automatically with a saving of fatigue and a striking gain in rapidity and accuracy. Indeed, simplified in this fashion, conduct attains a mechanical precision comparable with that of the insects. A succession of conscious movements becomes a chain of unconscious movements which executes the conscious idea of an action—or, still further generalized—of a course of behaviour. Ideas of particular actions drop out, and all that is necessary to initiate a series of muscular movements is a general idea of their character, and of the course which they should follow. So we execute such rapid movements as those of the tongue and lips in speaking and of the fingers in writing and playing the piano. Action of this description is called 'idiomotor': it is plainly of the kind which we have distinguished as *associative*, motor nerve-cells that are associated by familiarity of rhythm automatically stimulating one another in succession. It resembles primitive reflexes in being idea-less, and demonstrates that the process of memorial association is not confined to the mind. We remember because we have a nervous system, not because this system has culminated in a brain.

This memorial activity may not be simply repetitive, as it is, for instance, when a pianist executes a piece 'by heart'. It may be adaptive and suit itself to passing circumstances. We must, it seems, suppose that in this case the motor nerve-cells are also coupled with record-ideas of sensory impressions which have frequently attended their action. These would be called into activity by similar sense-impressions and would influence the automatic responses of the motor nerve-cells. So a pianist, playing to score, may follow the notes subconsciously.[6] In mounting the stairs one

[6] It has already been noted that Ferrier's researches led him to the

may be quite unaware of his steps.[7] Yet they will adapt themselves to the winding of the staircase. These automatic memorial associations, it is to be observed, are distinct from *habit.* For this is guided by a succession of ideas, and is conscious.

The general ideas which stimulate courses of action may be classed as of 'doing', 'saying', and 'making'. The actions and utterances which follow them may comprise elements that have been derived from either practical or manifestative movements or from both. But the distinction re-establishes itself in the *motives* by which actions and utterances are initiated; and behaviour may accordingly be distinguished as practical and manifestative. The motives of *doing* are practical when conduct is directed to the attainment of an object or of knowledge: they are manifestative when they express feeling, or when they take an imaginative form. It may be, however, that in our behaviour the practical and manifestative overlap. We use dinner napkins because it is dignified to do so, as well as because usage has demonstrated their convenience. And practical (in consciousness, *expectative*) behaviour may be tinged with imagination, as when it takes a romantic colour.

Through ideas of *saying* language has evolved. Words are muscular movements. The consonants are 'gestures' of the lips, tongue, and throat, which are in themselves as independent of sounds as are movements of the

conclusion that 'adapted actions such as intelligence would also dictate are capable of being called into play through the spinal cord, entirely without consciousness' (*Functions of the Brain*, p. 20). And Professor Sherrington writes to the same effect: 'Our individual experiences show how readily volitional acts, by repetition and practice, ultimately become actions involving neither attention nor even consciousness—create in fact habitual reflexes . . . testifying to a so-to-say primitive process akin to memory in the spinal cord' (Schafer's *Text-book of Physiology*, ii. 860).

[7] But he is unaware of them only so long as they excite no pleasure or pain. Through fatigue or a touch of rheumatism they force themselves on attention. So nervousness renders us at once self-conscious and inefficient—the fear that our words will not do us credit does not permit their utterance to pass out of the control of the conscious brain.

fingers. The vowels are formed by muscular action which drives wind from the lungs through the vocal chords—as through the reed of a clarinet—and through the cavity of the mouth, and determine the sounds so produced by varying the distance of the chords, and the shape and 'stopping' of the cavity. We are, then, provided with two instruments for the production of the vowel sounds, which may be likened respectively to a clarinet and an oboe. We use only the latter when whispering.

Language is primitively manifestative: the earliest utterances of infancy are expressions of feeling. And we shall see that emotion, in the form of imagination, has contributed very largely to the invention of words. But speech is of such immense utility that practical motives have strongly influenced the course of its development. Accordingly speech, like action, may be practical or manifestative: it may aim at achieving an expectation or may express a feeling. Its evolution will be discussed in a later chapter. But a few words of explanation may be offered in passing. Most words are meaningless symbols—originated by efforts of trial or imaginative creativeness—which have been associated in rhythmic coincidence with the ideas that they represent. In many cases, however, they have evolved from an appreciation of samenesses which has guided efforts of expression. In 'shriek' the voice imitates the shrieking, in 'sawing' the noise of the saw. Ideas of birds and quadrupeds are often linked to ideas of their utterances, and these are imitated by the names that are given them. Words signifying movements have developed in greater measure than may be suspected from an appreciation of samenesses between movements of the tongue and those of the limbs. In *pinch* and *jump*, for instance, the muscles of the lips and tongue reproduce in miniature the movements of pinching and jumping.

Ideas of *doing* and *saying* may evidently be summoned into thought by a familiarity of trait between them and the expectation or desire that is before us. If, for instance, we wish to arrive at a place *quickly*, we must progress *rapidly towards it*: the idea of day's *activities* suggests one of getting *up*. If our object is to write a *letter*, we must seek the *note-paper*; if to light the *fire*, we ask for a *match*. Accordingly ideas of action are commonly summoned through the intelligent appreciation of similarities or analogies. But they may be called up by memory as when we mechanically take our gloves when going for a walk. And, by repeated practice, intelligent connexions degrade themselves into purely memorial associations.

The evolution of ideas of *making* was a leap in advance. There is nothing in man's instinctive propensities[8] which could afford materials for forming them, and they must, therefore, have been invented by efforts of trial assisted by intelligence—that is to say, by an appreciation of samenesses. They again may be practical or emotional. In the former case they materialize an idea which has a useful consequence: in the latter they express an imagined idea—the beginning of Art. In both cases they have arisen from a sameness between the traits of a visual idea of an object or image and those of certain movements of the hands. There is, for example, a sameness between the outline of a sphere and a circular movement of the hand. A desire to make the sphere, or to draw it, may arise from imitative promptings, which we are about to discuss. An idea of the movement will present itself, and this will translate itself into action through its association with the motor nerve-cells that actuate its movements. But an inclination to imitate cannot develop into fabrication without an acute sensibility to samenesses of trait.

[8] If we except the ourang-utan's nest of branches, the only habitations made by quadrupeds are those of the rat tribe; and the uniformity of construction throughout the species shows that they are made under the guidance of ideas that are derived from reflexes. A similar conclusion can be drawn from the uniformity of the nests made by each species of bird.

We may conclude, then, that our actions and utterances arise from the invention of *ideas* of movements, and in this process (as is clearly shown by the discovery and improvement of mechanical appliances) analogy, or familiarity of trait, has played a determining part, although much has been owed to adventurous trial and fortunate accident. Methods of conduct and speech have evolved by steps that must have been slow and halting—the steps by which mankind has risen from the conditions of ape-like simplicity in which each man's life begins.

They are spread from man to man, and from generation to generation, by imitation. We imitate another by expressing in muscular movement visual or auditory ideas of action or speech which are afforded us by his behaviour. The visual or auditory idea calls up an idea of a movement of our own through an obvious sameness of trait which unites them: the idea so summoned expresses itself by associatively stimulating the motor nerve-cells that actuate the movement. Accordingly imitation, like fabrication, involves intelligence—but of a lower order, since the analogy that connects visual or auditory ideas with ideas of movement is less recondite. Fabrication is, indeed, an advanced form of imitation. One who fashions an object, invents a new appliance, or expresses an imaginative concept, has an idea before him which he imitates creatively by transforming traits of form into traits of movement; and his action may be styled *expressive* as opposed to *submissive* imitation.

Since ideas of action or speech tend to express themselves automatically in movements, and familiarity of trait may act subconsciously, we may imitate another without any conscious intention of doing so. This fact affords an explanation of the curious uniformity of behaviour—and even of feeling—which may infatuate a crowd. For, by auto-suggestion, conduct calls up the feelings which have been associated with it, and one who runs when others are panic-stricken, feels terror, although quite unaware of the danger that threatens.

By the educative effects of imitation the acquirements of each generation are transmitted to its successors so rapidly and completely that they appear to be inborn faculties. That they are really acquired accomplishments is proved by the striking differences between the manners and languages of different ages and different peoples. And we shall find reasons for concluding that a child brought up amidst brutish surroundings will not only speak no language, but will show no traces of human behaviour.

To recapitulate :

1. Muscular movements are primitively either instinctive reflexes, automatic expressions of feeling, or efforts of trial. The first are purposeful or practical, and, being instinctive, are stereotyped. The second being merely liberative, are more or less plastic. The third are free and therefore entirely plastic, and may take either a practical or an emotional character.

2. Movements become voluntary—can be controlled by the will and developed by intelligence—when they are initiated by ideas. These are formed from psychic expressions or efforts of trial which have been guided by a familiarity of trait between movement and expectation or desire, and have yielded favourable consequences. Efforts of trial develop into *practising*—the origin of skill.

3. Ideas of movements, unless inhibited, express themselves automatically by reversing a sequence of nervous association. In actual experience a movement and an idea of a movement coincidently follow the activity of the motor nerve-cell. The idea is, accordingly, associated with the motor nerve-cell in sequence, and calls it into action by a reversal of the sequence.

4. Ideas of movement are summoned into thought by familiarities of trait between them and the object in view, whether practical or manifestative.

5. Actions or utterances which have been practised in series become associated memorially into neuro-serial

sequences, which, once started, run automatically and unconsciously. Their course may be simply repetitive or may be varied by the effect of sensory impressions, which being memorially associated with certain movements, introduce new elements into the sequence.

6. Courses of action can, therefore, be initiated by general ideas of lines of behaviour.

7. The primitive distinction between practical and liberative movements subsists in the twofold character of behaviour as expectative and manifestative. Activities of each kind may be distinguished as of *doing*, *saying*, and *making*. The latter involves the expression in movement of visual or imagined ideas of form that possess analogies with movements.

8. Imitation involves the expression in action or speech of visual or auditory ideas of another's activity. Consequently it is closely allied to fabrication, which may, indeed, be styled expressive, as opposed to submissive, imitation.

9. The evolution of conduct and speech has been a slow process. But it has been accumulative, since by imitation each generation acquires the accomplishments of all the generations that have preceded it.

PART II

MOTIVES AND BEHAVIOUR

CHAPTER XI

THE APPETITES, EMOTIONS, AND WILL

MOTIVES and behaviour constitute what is generally called 'human nature', and we enter upon a less abstruse branch of our subject. We know far more about ourselves than we suppose, as is proved by the appropriate significance of our vocabulary.[1] But it is impossible to classify our feelings and actions, or to trace them to their causes, unless we realize, quite plainly, that our nature is threefold,[2] that life is a compromise between physical impulse, spiritual energy, and reflective calculation. Their rivalries give existence a dramatic character; and the motives and behaviour of others interest us so profoundly because they illustrate actual or possible experiences of our own.

On the physical plane we are affected by neural conditions which may be the consequences of organic changes, of other neural conditions, of sensory impressions, or of the two latter combined. Hunger, for instance, stimulates a search for food; favourable or unfavourable impressions urge movements of approach or recoil; the strange or difficult, efforts of escape or venture. If the neural conditions which attend these impulses, and their satisfaction or dissatisfaction, involve a great disturbance of equilibrium, neuro-serial reactions occur before the nerves settle down. This is so, in particular, with the neural states of which we are conscious as success and failure.

[1] Witness such words as 'thing' (appearance) for an idea, 'present' for the sensation of the moment, 'trait' for an attribute *drawn*, or isolated, from an idea, 'deliberate' for the weighments of choice, 'truth' for that which may be *trusted*.

[2] Σαρκία καὶ πνευμάτιον καὶ τὸ ἡγεμονικόν.—A little flesh, a little spirit, and that which governs. (Marcus Aurelius, *Med.* ii. 2.)

On the psychic plane these conditions are spiritualized. Favourable or unfavourable stimuli produce the psychic excitement of which we are conscious as pleasure and displeasure. Changes of nervous condition become the revulsions which we feel as joy and sorrow, pride and shame. Fear and love become 'passions'—that is to say, they are reinforced by an excitement which in itself is passive, and becomes active by being associated with physical promptings. Such promptings may also come from the reflective plane—from processes of thought which have the effect of extending our self-interest to others, as in the emotions of kindness and sympathy. Psychic excitement may, however, manifest an activity of its own. This is so when it takes the phase of *resistance*, a condition that is peculiar to animals whose nerves are, in some degree, concentrated. Zoophytes experience the repulsion of alarm, but manifest no sign of courage.

In all these phases emotional impulses are primitively involuntary: they involve no consideration of consequences. But voluntary motives evolve when psychic resistance stimulates the effort of venture which is so characteristic a feature of instinctive life. The effort may be assertive, tentative, or be guided by expectations of pleasurable or displeasurable consequences. So emotion evolves into the Will.

Emotion, as we have seen, has a further consequence. It stimulates imaginative reflection, in which thoughts take the concrete forms of images or fancies. Such are hopes and fears, which may compete with reasoned expectations in swaying the course of volition. And these images, in their turn, act as emotional stimuli, and reinforce the excitement which begets them.

In its elementary forms emotional excitement may be independent of the brain. The neuro-sensory causes of fear, for instance, touch the physical plane directly, and may be things as well as persons: the influence of the persons by whom sexual and maternal love are aroused is similarly experienced instinctively. But affection, faith, respect and pity, jealousy and scorn, anger and revenge involve the entry of *ideas* into the neuro-sensory process—that is to say, they cannot be experienced without the intervention of a brain. They are only aroused by impressions of another person's behaviour, when these are associated with certain ideas of his feelings or relations towards us. Consequently they cannot be aroused by inanimate stimuli. We perceive the feelings of another, not by sensory impressions, but through recollections (record-ideas) of feelings which are coincident with ideas of such words as he utters or of such expressions and actions as he displays. If an idea of another's feeling is called up in this fashion, it stimulates us emotionally by exciting us in sympathy with, or in antagonism to, the feelings which constitute our own personality. We are, accordingly, angry with another because we have an idea that he is disposed antagonistically towards us. We may, of course, be *irritated* by contrariety in the inanimate, for this causes nervous depression which becomes irritation when infused with resistance. When the misery of severe illness breaks into ill-temper, the patient is gathering strength. We share this antagonistic impulse with brainless creatures, such as insects, which possess a concentrated nervous system. But the emotion of anger requires as its stimulus an *idea* of a hostile personality which is given by the brain, and, if no such idea presents itself, anger is not aroused. In the course of a football match injuries are given and received in good temper because they are not inflicted with angry intent.[3] It follows that we cannot be angry with the inanimate. A child may revenge itself upon an obstructive table, but this is because it regards the table animistically. A collar-stud which *angers* us is figured as 'possessed'.

[3] This association of ideas influences the law. An action is not criminal, unless it is committed with criminal intent.

Consequently, we can be emotionally excited by an idea of ourselves. We possess three characters according as we yield to temptation, resist it, and are conscious of the result. There is a standing antagonism between submission and resistance: it follows that we feel triumphant if we overcome the idea of yielding, and experience shame, and the revengeful feeling of remorse if we are overcome by it. But when no antagonism is involved, as in undeserved misfortune, we can unaffectedly pity ourselves.

Moreover, our emotional life is profoundly affected by the manner in which emotional stimuli—that is to say, the persons who affect us—are classified by the brain. Other persons are unified with, differentiated from, or contrarified to us, when in some distinctive trait they are the same as, different from, or contrary to ourselves. In this process the brain acts exactly as in perception. Unification, differentiation, or contrarification may be in kind, or (imaginatively) in personality. We identify with ourselves in kind those who possess kind-traits similar to our own: thus a soldier identifies himself with another soldier, an Englishman with another Englishman, a Liberal in politics with another Liberal. A far closer identification is that in personality. In this case we are 'at one' with the other: his joy is our joy, his sorrow our sorrow. In a word, we *sympathize* with him. By such unifications, differentiations, and contrarifications the influence of physical stimuli may be completely reversed. From a comrade one may receive an injury without resentment. On the other hand, favours that are offered us by an enemy, or even by a stranger, may annoy us exceedingly.

Through the sympathy which follows unification in kind or personality, we are affected by benefits and injuries which we do not ourselves experience, but are offered by another to a third. If neither is friend or foe, we can contemplate their relations with dispassionate, neutral, curiosity. But, if either of them is our friend or foe, our emotions are aroused—on behalf of the friend with whom we identify ourselves, and against the foe whose personality is the contrary of our own.

Further, the brain heightens the effect of emotional stimuli by imaginatively embellishing them. Hence danger may become more fearful than it is in reality: one whom we love is endowed with fanciful excellencies, one whom we hate, with imaginary sins. We accordingly exaggerate both the good and the evil of life, and find it difficult to take a temperate view of its emotional incidents.

Since ideas, or idea-traits, can bring about neural states with which they have been associated in experience, we can understand why emotions can be excited by recollections of stimuli, by ideas of them that are suggested by others, and even by imagined stimuli. Any emotion may be stimulated imaginatively. Children are touched by emotion through their toys according to the feelings with which they invest them in idea. The characters of fiction and the drama recall ideas of emotional states which very commonly excite actual emotions in ourselves, if we 'take sides' with or against them. We shall not wonder at the emotional power of imagined stimuli, if we remember how extraordinarily imagination can influence the functioning of the organs of the body.

The origin of the appetites and the emotions has already been discussed in chapters ii and iii, but we may review them in the light of the more advanced conclusions at which we have now arrived. The appetites are apparently excited by *organic* stimuli. Food, drink, and rest are physical necessities, and, when they press us urgently, naught but fear can withstand the urgency of satisfying them, although, as we shall see, ideas of pride or honour may nearly succeed in doing so. The things which satisfy them are associated with the pleasure of satisfaction. Thus we like our food, quite apart from its taste, and the earth-eating Indians can appreciate a diet which can afford them no more than a sense of repletion. The promptings of lust, when associated with another, become sexual love—an emotion which is spiritualized by the psychic feelings which it excites, and is glorified and stabilized by imagination and sympathy. We recognize, nevertheless, that it is fundamentally instinctive and may be quite 'unreasonable'. Maternal love also arises from an instinctive craving, and for this reason is indifferent to all ideas of consequences.

Next of the physical conditions of fear and dread that are caused by the *neuro-sensory* effect of the strange, the injurious, and the antagonistic. Under psychic resistance fear becomes courage, activating a movement of advance in place of one of escape; depression or dread becomes irritation or anger. In these, resistant expansion does not wholly neutralize contraction, because the stimulus, being antagonism, or the idea of it, is a continuous irritant. Anger is, therefore, a mixed emotion. It prompts an imitative effort of retaliation or revenge, the idea of which is suggested by the injury or opposition which we have ourselves suffered, but may be modified by experiences of consequences. If the opposing personality is unified with ourselves, retaliation, in the form of punishment, may lose all bitterness. A mother can correct her child without anger. A judge also condemns dispassionately. But this is because he is a third party, and has suffered no personal loss or injury.

The nervous relief which is the neuro-serial consequence of another's protection becomes emotional as the confidence of faith. Since his motive is in accord with our own, he is unified in kind with us, as our representative or chief; and feelings of confidence and loyalty are associated with the idea of him. But if he is contrarified as an enemy, we may resent his services while taking advantage of them. Accordingly, faith arises as a nervous reaction in which a contraction of fear is followed by an expansion of relief. It includes a feeling of respect, for protection involves power, the idea of which stimulates associatively an expansion that is followed by a neuro-serial contraction. This infuses faith with an element of gravity which renders it peaceful rather than pleasurable. Faith is active, because if we trust another we follow him, or imitate him; and imitation that is infused with faith is obedience. Our acceptance of his authority generally reposes upon an *idea* of his protective power, not upon actual experience of it: accordingly, we may be disillusioned by circumstances, lose confidence, and revolt. We may place self-confident trust in our own power: this can only come from experience, and can never be as strong as faith in another.

The love that is termed 'affection' or 'attachment' also arises as a feeling of pleasurable relief—that of *satisfaction*—due to the helpfulness of another, which, being associated with him as its cause, will be stimulated associatively by an idea of him. Pleasure that is consequentially associated with another is affection. It is reinforced by the attendant psychic revulsion of joy: love and joy are, we feel, akin. Children regard with affection[4] those who minister to their needs. We term this affection, depreciatively, 'cupboard love'. But we do not reproach ourselves with loving our benefactors. On the other hand we hate that which causes a revulsion of sorrow, as by suddenly withdrawing the expected satisfaction of an appetite. Love and hate are active emotions because the conduct which excites them, being in our interests or opposed to them, naturally provokes the brain to unify with, or contrarify to ourselves the persons from whom it proceeded.

Respect and pity, jealousy and scorn are similar and yet different—the latter pair being the precise opposites of the former—and show very strikingly the effect of the mental upon the emotional. The evolutionary origin

[4] A dog loves him who feeds it, and a robin will contract an affection for the gardener whose spade turns up its diet of worms.

of respect and pity can be traced to reactions from neural states of success and failure, which, being conditions of extreme inequilibrium, are followed by nervous oscillations before balance is restored. Respect represents a contraction from the expansion of success. It is a neuro-serial consequence—a *shrinking*, akin to fear, and when associated with another, it prompts us to stand aloof and expresses itself by obeisance. Pity, on the other hand, is an expansion from the contraction of failure: accordingly it draws us towards one with whom the failure is associated, and expresses itself in acts of helpfulness. Conditions of expansion or contraction are stimulated by success or failure in idea as well as in actuality; and, since power and superiority are idea-traits of success, and weakness and inferiority idea-traits of failure, ideas of these also act as stimuli. They are associated by the brain with ideas of personality. If our own personality is associated with ideas of superiority or of inferiority, we regard ourselves with respect or pity; and, if these ideas are associated with another, our respect or pity goes out to him, and can be aroused at any moment by an idea of him. But this is only so if we unify him with ourselves [5] in kind, and do not attribute to him a hostile attitude. If he is figured as different or contrary in kind, respect is infused with antagonism and becomes jealousy: pity, similarly tainted, becomes scorn. And, since the good are favourable, and the bad unfavourable, we feel pity for the one, scorn for the other.

Accordingly, jealousy is a state of nervous contraction that, under the influence of the brain, is infused with antagonistic resistance. It becomes active as hate and prompts an angry effort to injure. It is the most subtle [6] of all our antagonistic emotions, insidiously

[5] As the Good Samaritan identified himself with the man who fell among thieves. If one is identified with us *in personality*, we do not merely pity him in his distress, but suffer with him.

[6] Since dress affords pride, differences may excite jealousy. Hence the common practice of dressing young sisters alike.

setting us against every one who is not our inferior, if he is in no way identified with ourselves. We cannot be jealous of children: they do not irritate us by any suggestions of rivalry, and there are few who have not smiles for them. 'Little' is a term of endearment because it waves jealousy aside. Nor are we jealous of the dead, since they no longer compete with us, and we may praise them—indeed overpraise them—unaffectedly. But, for the rest, the success of another irritates us, unless we unify him with ourselves in kind or in personality.

Finally, of resistance—the one psychic emotion which is active in itself, and does not depend for its motive power upon the physical or reflective plane. From it volition has evolved. But, primarily, it is the opposition that is offered by psychic energy to interruption, and may, therefore, be classed as a *neuro-serial* consequence. Accordingly in a mood of expansion we resist the effects of annoyances, and, when irritated, we object to being pleased. Expansion and contraction—that is to say, pleasure and displeasure—are mutually contradictory: one interferes with the other. There is a similar 'incompatibility' between physical and psychic impressions—those of our environment and those of ourselves—and accordingly the psychic is disposed to resist the interruption of the physical. We may detect the influence of this tendency in ourselves. At moments of emotional stress offers of food and other comforts are resented: we feel 'above' them. If we examine the most primitive rules of morality, we find that they are concerned with resistances to physical promptings and stimuli, which are imaginatively figured as 'polluting' the 'purity' of psychic feelings and thoughts. There can be no question that morality involves resistance; and since resistance is psychic, and therefore emotional, the moral contains an emotional element. We can feel that this is so. The emotion is, it is true, of a very different quality to that of fear or love. For it contains no physical element. Like courage, it is resistant, and opposes itself to instinct. It is, in fact, a form of *willing*.

More obvious phases of resistance are those involved in curiosity, industry, or perseverance, and in emulation, which oppose themselves to contractive excitement caused by the strange, and by ideas of obstruction, or antagonism, that is offered to us by things or persons. These resistances are primarily involuntary and independent of consciousness. They are suggested by ideas; but the ideas affect us subconsciously. They are, however, followed by *conscious* revulsions of pride or shame, according as they succeed or fail; and these are pleasurable or displeasurable *consequences* that incite us to practise resistance voluntarily for its own sake. We *will* to resist, and our effort becomes deliberate, instead of automatic.

The Will, as ordinarily so called, is individualized by us as a single faculty. But it assumes very different phases, which we have distinguished in chapter iii as assertive, tentative, and selective. We can infer—and, indeed, *feel*—that volition is effort that is stimulated by psychic resistance. The effort may be random, so testifying to its origin in the struggle of venture which is the elementary physical response to difficulty. Being stimulated by resistance, and resistance running counter to the effect of stimuli, volition appears to be spontaneous—a consequence without a cause. But resistance is itself stimulated by the incompatible—by that which it resists. Without nervous depression there could be no resisting irritation. The various phases of willing are resistant efforts which are evoked and led in particular directions by different kinds of stimuli.

Assertive volition may be 'wilful' contrariety, expressing irritation that has been stimulated by the unfavourable, and resisting its effect. It may be an automatic revolt against the regulated procession of physical and mental sequences that are incompatible with the vivacity of spiritual energy; or a conscious protest, stimulated associatively by an idea of antagonism or of effort. Accordingly assertive volition may prompt us to any effort of which an idea occurs to us. The neural state of effort can apparently substitute itself for any neural state that leads to muscular action, and so can produce its consequences. It is in this fashion that attention comes under voluntary control. It is primitively a neural state on the physical plane, which is followed by adjustments of the sensory organs that exclude all objects but one from perception. In 'forced attention' these adjustments are actuated by the will, stimulated associatively by an idea of effort. They are of the nature of muscular movements. But, movement and utterance apart, we cannot will ourselves into neural conditions that involve physical elements. We can will to be courageous, inquisitive, industrious, or moral, because these are purely psychic conditions, and are, indeed, subconscious phases of volition. But no effort will make us loving, angry, pitiful, or jealous, since these emotions are rooted in the physical plane. We may, it is true, force ourselves into them by imagining stimuli. One can, for instance, think himself into anger by fancying injuries, the traits of which are derived from injuries actually experienced. It is only in this fashion that we can 'will to believe'. The brain, however, stands apart from the physical nervous system, and we can, therefore, influence its working by assertive volition. We can force attentiveness in reflection by a 'strain' of which we are well aware when we concentrate thought upon an uninteresting subject: we can *compel* our thoughts, can summon ideas capriciously, can connect one idea with another, or reverse a connexion. And, if the ideas so summoned are stimuli, we may indirectly produce in our physical and psychic selves neural states which the will could not of itself bring about.

Tentative volition is impelled by doubt or curiosity

and manifests itself in experiment—an effort which may be made at random or be led by expectations of consequences. We shall see how greatly we are indebted to it for the evolution of our methods of conduct and speech. These, when once invented, are acquired by imitating others, and become habitual. But we are influenced by volition in acquiring them. 'Learning' is imitation that is *willed*.

When volition involves *choice*, it is selective. A number of possibilities occur to us, and their contradictions bring about the mental confusion of hesitation. Willing resists this confusion, and stimulates an effort which, being undirected by instinct, is led by expectations—by possibilities that are *appetitively* anticipated—and *chooses* the expectation which is appraised in memory as the most advantageous or least disadvantageous. It is instinctive to approach the favourable and recoil from the unfavourable, and accordingly selective volition is, in fact, led by instinct.

Amongst the expected consequences from which we choose, the pleasure of pride and the displeasure of shame weigh very heavily. They may force us to contend against our appetites, against fear, against love itself. And since pride and shame may come to us through the attitude of others, their opinion of our conduct is an attracting or repelling consequence. How many elopements have been chilled off by thought of 'what the world will say'! Moreover, the pride or shame may not be our own but of one with whom we sympathize. A man will risk much that a comrade may be proud of him. Accordingly pride and shame, as expectations, are incessantly opposing themselves to natural physical gratifications. Fasting, even for some weeks, is not impossible to one who wishes to gain notoriety thereby: to St. Simeon Stylites, on his column, complete rest was impossible: holiness has, in some measure, used celibacy as its advertisement. But the will, while bridling conduct, may be unable to stifle the agonized cravings which are reflected so insistently in the prayers and meditations of the cloister and the hermitage.

The possibilities which guide selective volition may not be inferred from experience. They may be hopes and fears—the creatures of the imagination. And imagination affects our willing when we choose self-consciously, romantically, or idealistically. In all these cases we follow an image because it pleases us. The self-conscious man pictures himself with pleasure as triumphing over temptation. The ideal of liberty is enchanting because it implies self-power. Sympathy and charity are attractive because they involve extensions of self. But in following them we do not appear to be pursuing pleasure, since we realize them, not by attainment, but by expressing them in our conduct. We may, for instance, dispose ourselves sympathetically towards others by means of an ideal Unity. This will lead us to notice samenesses between others and ourselves, which will have the effect of imaginatively unifying them with us.

If, however, selective volition is led by pleasure or displeasure, why do we think of it as 'free'? Firstly because it may revert to the random, as when one settles a difficulty by 'tossing up'. Secondly, because it is swayed by the influence of a liking, as opposed to a command, and one who follows a liking is *free*. Thirdly, because deliberation involves resistance to all the stimuli that present themselves up to the moment of choosing, and to resist a stimulus appears to involve freedom from its control. But, none the less, we are ultimately drawn by the most attractive of the possibilities that are before us. And in our heart of hearts we know that 'we have no power of ourselves to help ourselves'.

The development of selective volition is clearly the main cause of the progressive complexity of pursuits that is termed 'civilization'. For the pleasurable or displeasurable expectations upon which it depends can, as we shall see, be multiplied indefinitely. They are evolved through the action of intelligence from consequences that have been learnt or actually experienced —through the isolation of traits and their generalization. Or they may be ideals, formed by the imaginative personification of nervous conditions. The lower animals are but poorly endowed with the mental susceptibilities upon which these processes depend, and, consequently, selective volition plays but a small part in their lives. A dog, or cock-bird, for instance, is triumphant in victory, dejected in defeat. But, being unable to differentiate (or isolate) traits and generalize them, it cannot associate pride or shame with symbols. Hence no evolution of ambitions is possible. In man, volition appears to overshadow impulse, because the influence of the will involves reflection, and is therefore appreciated more clearly by us than that of the emotions. We see that the relations of the citizens of a State to one another, and the international relations of different States, are determined very largely by considerations of advantages and disadvantages—by goods or evils which may be phases of honour or dishonour, pride or shame, but commonly take a pecuniary or commercial form, since profit and loss are easily reckoned in money values. We do not realize so vividly that these voluntary pursuits and avoidances owe their effectiveness to the support which they receive from emotional energies. Without courage no people can be victorious—and anger contributes largely to the spirit of victory. Without the affections, there would be no ties of family and friendship; without faith and loyalty, no unifying government; without moral resistance there can be no order in social life; without industry and emulation no material progress. Before, then, passing to the influence of calculating volition on human affairs, we should give more detailed consideration to these great emotional forces.

In the physical, emotional, and volitional motives

which we have been discussing, the strands of instinctive, psychic, and reflective nervous activity are so intimately interwoven that it is difficult to summarize our conclusions briefly and clearly except through an analytical comparison. This is made in the tabular synopsis [7] annexed. But, to complete this chapter, we must refer to the bodily changes that accompany the emotions, and mark themselves in our *feelings* of them; and to the striking differences in emotional susceptibility which distinguish one person from another.

Acute fear deranges the action of the heart, lungs, and digestive organs, and relaxes the control of many muscles. Courage is attended by a quickening of firm pulsation and respiration. We can feel that anger, scorn, and jealousy affect us bodily. It might seem that the internal effects of an emotion would lie beyond the reach of experiment. But physiological science has succeeded in detecting a very remarkable glandular [8] activity which attends the conversion of fear into anger, and heightens anger into rage. A peculiar secretion (adrenin) is poured into the blood by glands that are situated above the kidneys, which has the effect of increasing the sugar in the blood, and so invigorating the muscles and strengthening the action of the heart and lungs. It has a further serviceable action: it facilitates the coagulation of the blood, and is, so far, an antidote to injuries which are received in angry conflict. Sharp psychic revulsions, whether expansive or contractive, affect the circulation, causing the changes of pressure that are manifested in blushing. Peculiar physical [9] conditions accompany the emotions of sexual and maternal love. Do they not also attend the less passionate affectionate emotions? It is not known.

[7] Note C.

[8] The experimental evidence is fully stated in Cannon's *Bodily Changes in Pain, Hunger, Fear, and Rage* (Appleton, New York). See also a paper by Dr. F. H. Kooy on 'Hyperglycaemia in Mental Disorders', published in *Brain*, vol. xlii, Part III.

[9] Sappho's celebrated description of the love-fever is pronounced

But pity, and the feeling of comradeship, certainly appear to involve physical conditions [10] of which we are conscious as a 'glow' of affectionate interest.

Susceptibility to the appetites and the emotions varies greatly from one individual to another. There is not a nurse who would deny that from earliest infancy children differ markedly in emotional disposition—that some are more courageous or more wilful than others; that some are instinctively disposed to affection, others to jealousy. It appears that sensibility to emotional stimuli varies, as does sensibility to the aesthetic pleasures of sensation. We start differently equipped by Nature. But there is the great counterbalancing fact that the pleasures and displeasures which compete with our primitive motives may be strengthened by practice so as to gain upon their instinctive antagonists. For practice, as is well known, facilitates the response to a stimulus by strengthening the association of nerve-cells and rendering their co-ordination habitual. By yielding to a stimulus we increase its power over us.

Practice may be forced upon us by our environment. A life of hardship is generally conducive to courage and self-control. But our environment includes the

by Longinus to be as true to nature as sublime in expression; and those who are warm-blooded may endorse this opinion:

> God-like is the fortune of him who sits close to thee,
> Catching thy whispers and watching thy smiles.
> A glimpse of thee sets my heart bounding in ecstasy:
> My voice fails; my tongue 's tied, bereft of its wiles.
> I glow with swift heat; my eyes cloud; my ears sing;
> I am damped by slow chill; and with quivering breath,
> As pale as the moon-weed, I feel my soul drifting
> And fading away towards the blankness of death.

[10] We are obscurely conscious of internal emotional conditions, and have endeavoured to locate them. Thus courage, compassion, and jealousy have been associated with conditions of the liver, the bowels, and the spleen: the heart has been supposed to be a spring of affection, and has given shape to the most popular tokens of love.

suggestive influence of others: their actions and words call up ideas which may affect us as strongly as material stimuli. So one who is more inclined than his fellows to the kindly emotions can, by his example or his precepts, effect radical changes in the susceptibilities of those who respect and believe in him.

NOTE C

Nervous planes contributing.	*Emotion.*	*Stimuli.*[11]	
The physical reinforced by the psychic, and the mental (memory).	Fear.	The strange, dangerous, or antagonistic in things or in persons.	The strange would not be appreciated unless the familiar was established by nervous association (memory).
	Sexual love.	Sexual cravings centred upon one of the opposite sex.	Arise as instinctive impulses, but are spiritualized by psychic and directed by mental influences.
	Maternal love.	Maternal cravings centred upon a child.	
The physical influenced by the mental (intelligence) and reinforced by the psychic.	The love of 'attachment'.	Satisfaction, or helpfulness, by one who is favourable in intent or unified in kind.	Arises as a nervous condition which is associated memorially with another.
	Faith.	Protection by one who is favourable in intent or unified in kind.	Arises as a nervous revulsion which is associated with another.
	Respect and Pity.	Power and weakness in one who is favourable in intent or unified in kind.	Arise as nervous reactions, the cause of which is associated with another.
The physical, led by the mental (intelligence) and reinforced by the psychic.	Loving-kindness, Sympathy, and their contraries.	Samenesses and differences leading to unifications, differentiations, and contrarifications in kind or personality.	Arise out of mental processes which extend to another the self-love of egotism, or contrarify another with self antagonistically. Can be stimulated voluntarily through ideals.
The physical, influenced by the mental (intelligence) and arousing psychic resistance.	Anger, Hate, Revenge.	Injury or opposition from one who is unfavourable in intent, or differentiated in kind.	Arise as nervous conditions which are associated with another.
	Jealousy and Scorn.	Power and weakness in one who is unfavourable in intent, or differentiated in kind.	Arise as nervous reactions, infused with antagonism by the brain, which are associated with another.

Psychic resistance stimulating an effort **that** is directed **by ideas.**	Curiosity, Courage, Emulation, Industry, Morality.	Conditions of doubt, alarm, or confusion caused by strangeness, danger, antagonism, or incompatibility; or, associatively, an idea of contrariety or antagonism.	Can be stimulated voluntarily by expectations of consequences (pride and shame) and through ideals.
Psychic re**sistance** stimu**lating** an im**patient** effort **that** follows **mental** sugges**tion.**	Assertive Volition.	Conditions of discordance caused by antagonism or incompatibility.	May be petulant, or capricious.
The psychic **resistance** of cu**riosity**, stimu**lating** an effort, **guided** by men**tal** expectations.	Tentative Volition.	Doubt caused by the strange, unfamiliar, or difficult.	Manifests itself in experiment.
Psychic re**sistance**, giving **scope** to an **effort** of pursuit **or** avoidance, **stimulated** by **mental** expectations.	Selective Volition.	Hesitation resulting from competing expectations.	Hesitation is stilled by resistant effort: choice follows. It may be self-conscious, romantic, or idealistic.

[11] **These stimuli need not be actually experienced: they may be recollected, or suggested in idea by another, and may even be imaginary, as are those presented in fiction, at the theatre, or through symbols. And any emotion may be experienced imaginatively in sympathy with one who is experiencing it.**

CHAPTER XII

FEAR, COURAGE, AND ANGER

FEAR is so masterful a nervous condition because it is the safeguard of life: an animal which was absolutely fearless would not have many days of existence. It is, as we have seen, the psychic counterpart of physical conditions of shock, which may be caused by lifeless as well as by living stimuli. We may particularize *alarm* as the shock of the strange, *fright* as that of the dangerous or injurious in idea, *dread* as that of the difficult or antagonistic. The effect of the dangerous is reinforced if it is also strange. Fear may then become terror, as was illustrated by the more timid classes of the population during the early days of air-raids, when they were a strange experience. Troops must be well seasoned to stand the shock of a surprise, and consequently from the beginning of history ambuscades have been an effective feature in military tactics. On the other hand, when the dangerous becomes familiar it is less formidable. If troops have become accustomed to a new instrument of destruction, such as asphyxiating gas, they will face it composedly. In towns which have been long bombarded, children will play while the shells are falling.

We have seen that, by a process of auto-suggestion, conduct that has been associated with an emotion recalls the emotion. This association is illustrated very strikingly by the onset of panic. One man begins to run because he involuntarily imitates another who is running, and, as he runs, his fear increases. Indeed by the mere act of running one may feel that he is pursued—a fancy which is evidently experienced by a dog that races in play round and round a lawn.

The shock of the difficult or antagonistic becomes emotional as dread, or nervousness, a feeling of perturbation that may be caused by the opposition of the inanimate or by the clash of another's feelings or motives with our own. It renders the timid shy and awkward in society because they may fail to shine in it; and underlies 'stage-fright' and the hesitation which so often attends public speaking. These are, however, trivial phases of a nervous condition which, if acute and stimulated by living antagonism, often leads to an outburst of anger.

An apprehension of the strange or dangerous may be followed by feelings of admiration if it manifests excellence or power. Fear becomes awe, a complex emotion in which nervous contraction is attended by an expansion that relieves it of its unpleasantness. We stand in awe of a violent thunderstorm or an avalanche, or when we contemplate the supernatural. We may experience the feeling when we enter into the strange immensity of a Gothic cathedral. The relieving expansion may not be of admiration but of reassurement, arising from the fact that what is fearful in idea is not so in actuality. This is the charm of the mysterious, which renders, for instance, a ghost-story attractive. Few of the children who are entranced by it would care to convert its suggestions into actualities—to explore, for instance, a churchyard at night. And when the mysterious ceases to be strange it loses its hold of us; this is so even with the words which we employ to express its influence. 'Awfully' comes to mean simply 'very'. The mysteries of Nature confront us daily. We are not greatly impressed by them. Familiarity can even breed contempt.

By resistance fear is transformed into courage, and depression into the irritation of anger. The former may be a wholly expansive condition; in the latter there is a strong element of contraction because it is excited by the idea of a hostile (or *contrary*) motive which obsesses us and maintains conditions of contractive excitement. If there is no idea of an opposing personality, our feeling is one of antagonistic irritation: we may be enraged by the 'malevolence of inanimate things', but only by endowing them with malevolence. The resistance of courage or irritation may be quite involuntary. It is clear that some insects are excited by them: the courage of the fly is proverbial. A nervous state of resistant excitement can exist without a *feeling* of it: dogs that have been deprived of their brains will bristle and growl if roughly handled. Courage and irritation can, then, be evoked automatically by stimuli the effect of which excites resistance. Courage may also, of course, be stimulated by a conscious effort of will. But we will consider, in the first place, its involuntary phases.

Courage, being the resistance which is offered by psychic conditions of expansion, will depend for its automatic arising upon the elasticity of one's psychic energy and the intensity of the expansive excitement of the moment. Elasticity, we may infer, is increased by cold and dryness, and is lessened by damp heat. We can feel the effect of these atmospheric conditions upon our spirits, and we recognize their permanent influence in distributing obstinate hardihood and evanescent bravery amongst the nations of mankind. Some are endowed with 'bull-dog' pluck and rallying power; others with a brilliant, but passing *élan*. The peoples of Northern Europe and America recover from the effects of defeat more rapidly than do Southerners and Asiatics; and this peculiarity has had an immense influence upon history, since it has given them success in overrunning the world. For in warlike enterprises both sides must from time to time suffer defeats: the ultimate victory is to those who can most rapidly recover[1]

[1] A complete loss of automatic resistant elasticity characterizes the nervous collapse of 'shell-shock'. Conscious resistance remains possible. But it involves an intolerable strain.

from the effect of reverses—a truth which is constantly illustrated in the pages of history. The nations of the well-watered southern lands, however brave has been their initial resistance, have invariably succumbed to invaders who have come from the deserts, or from colder and less enervating climes.

Psychic elasticity may, however, undoubtedly be increased by practice. Courageous resistance is strengthened by the training of the polo and football ground, and by the self-control that is required in the sterner forms of sport. A life of hardship, involving continued effort, fosters resisting power, whilst luxury tends to enfeeble it. Each time, therefore, that we resist the shock of fear, or dare it in bravado, our sensitiveness to it decreases; each time that we shrink from danger, take shelter from it, or seek the protection of others, we increase the difficulty of meeting it in the future. Discipline, as is well known, facilitates courageous conduct. It acts, however, through the force of habit. Feelings are so closely associated with behaviour that actions [2] repeated by habit recall the feelings that in the past have inspired them.

Elasticity apart, the resistance of courage is reinforced by anything that adds to the intensity of expansive psychic excitement. This may be the effect of a physical stimulus. An army's fighting strength is materially dependent upon the satisfaction of its stomach, and if troops are ill-fed, they will lose their dash, unless they are excited by strong emotion. Music and alcohol are exhilarants. Hence the military value of paeans, battle songs, and regimental bands, of the skirl of the pipes when the moment arrives to go over the parapet; also of the final tot of rum, and the glass of cherry brandy which is as 'jumping powder' to the hunting man on a depressing morning. The stimuli may be emotional. Angry antagonism and its climax in rage are phases of

[2] Or even utterances, as is illustrated by the effect of the cries of rage which have been prescribed as a feature of bayonet exercise.

resistance, and may stifle all qualms of nervousness. A mood of exhilaration is still more inspiring, and courage becomes effortless when one is moved by love, pity, comradeship, loyalty, or faith. In defence of her child a mother becomes indifferent to danger. Love[3] will infuse courage into the most timid of men. Pity for a wounded comrade will not only nerve the soldier to rescue him, cost what it may : it inspires the stretcher-bearer with that highest courage which owes nothing to personal pride or the hope of glory. Man fights well when he is surrounded by his friends, or feels that he is defending his home, when, that is to say, he is stirred by emotions of comradeship or affection ; most stubbornly of all when he is devoted to his leader in faithful obedience. This gave the martyrs their astonishing courage.

Finally, the reinforcing stimulus may be ideal. Regiments may be inspired with courage by speeches on the battlefield—still more effectually by the courageous example of their leader. In time of war the suggestion of such ideas as may infuse courage is the most urgent occupation of statesmen who mean to lead a nation to victory through the discouragements of a protracted conflict. For courage is unstable, being less elementary than fear. The *moral* of an army, or of a nation at war, is a delicate plant, and requires to be carefully fostered—not only by appeals to patriotism or revenge, but by reassuring accounts of the enemy's decline. The people must be restrung elastically, as it were, at frequent intervals by the dexterous propaganda of politicians and journalists. And their appeals are most effective when they can recall ideas of affection or reverence, of comradeship, pity, loyalty, and religion, since these are more enduring stimulants than malevolence. The spirit of revenge gives courage, since it originates in the stimulus of actual injury. But appeals

[3] The breath of spring, which sets birds mating, fills them at the same time with quarrelsome ardour.

to jealousy are less effective, because this may be a passive feeling, unless it calls to mind an injury that has been actually sustained. A Hymn of Hate is a less effective stimulant than a glass of whiskey.

We pass to voluntary courage. Since success and failure have consequences, in pride and shame, that are keenly pleasurable and displeasurable, we can *will* to be courageous by choosing to seek the one or avoid the other. We can brace ourselves by thoughts of the pride of victory: we may be steadied by the fear of shame or of self-reproach. Steady courage is aptly termed 'moral', since it is urged by that fear of disapprobation which is the bulwark of morality. It may not extinguish fear. The man who dares because he is afraid to be a coward is not cheered by the light-heartedness of automatic or self-assertive audacity. But he resists. And one may stand fast by choosing the least of two evils—the risk of combat in preference to that of being shot for cowardice. This may keep men in the ranks. But it will not inspire them with the fighting spirit.

And we may choose to be courageous under the influence of a pleasurable ideal, which is pursued by being expressed in our conduct. *Noblesse oblige* is such an ideal: it is attractive because it implies an idea of pride. There is the same implication in the ideal of patriotism, since our country is unified with ourselves. Magnanimity is an ideal of transcendental refinement, since it implies the power that is involved in resistance quite apart from its consequences—that is to say in *ascetic* resistance: it represents spiritual energy in itself, and is the most enduring support of courage in that it is actually stimulated by difficulties. Such was the stoical courage of Socrates and Epaminondas—perhaps the most magnanimous of classical exemplars. Ideals may be flashed upon us by others. Patriotism is, of course, one of the most effective themes of persuasive oratory. Magnanimity is less commonly invoked; and this appears to be characteristic of our times, in which asceticism is fading under the shadow of 'economic interests'.

Moreover, courage, being a condition of expansive psychic excitement, is pleasurable independently of its consequences, and is therefore an attractive ideal in itself. It is not, like anger, sullied by irritation. Consequently, our hearts go out to deeds of bravery, and there is no human trait which we admire more sincerely. We may literally speak of the 'happy warrior', and need not wonder that Nietzsche, on a review of human feelings, should have concluded that courage is best of all.

Turning now to Anger, it may be observed that it is easier to be angry than to be courageous, because anger comes in great measure from yielding to an irritating stimulus, and is resistant only in becoming active. It automatically manifests its activity in retaliation or revenge; and since this angers its object, and provokes him to exact vengeance in his turn, society must have been distracted by endless vendettas until the discovery was made that the punishment of the aggressor by a third party not only checked aggression, but satisfied the vengeance of the aggrieved. Legal punishment is a double antidote: it checks the retaliation of wrong-doing as well as wrong-doing itself. Justice is the Goddess of Peace. If seems evident, therefore, that international peace is a dream of Utopia unless a super-national authority can be established with power actually to punish aggressiveness, since this alone will satisfy the desire for vengeance. A heavy fine might suffice. For one of the most striking lessons taught by history is that even a deadly offence to one's honour—an injury which in years not long past could only be expiated in personal combat—can be revenged by the legal infliction of a money penalty: indeed, a claim for money damages is a prominent feature in suits for divorce. It is to be observed that, since we are emotionally excited by ideas of our own personality, vengeance may be aimed at oneself. Tortured by self-caused shame, one experiences the agony of *remorse*, which may force a man to retaliate upon himself—by yielding himself up to justice, or even by suicide.

There is a remarkable phase of anger termed 'indignation' which arises when our sense of *justice* is shocked. This concept, as we have seen, is generated by the unvarying regularity of our nervous reactions, which leads us to expect a similar order in the world outside us. The resistances of virtue should be rewarded —that is to say—respected: the yieldings of vice should be punished—that is to say—contemned. Consequences that are the contrary of these are 'unworthy'; they arouse our indignation because they are antagonistic to one of the strongest of our expectations —so strong, indeed, that the hope of a reign of justice is one of the most heartening of the consolations which religion can afford.

We cannot *will* to be angry, since anger, although including the psychic element of resistance, is fundamentally a physical condition of nervous depression, whereas courage, running contrary to—and annulling— the nervous contraction out of which it arose, can be dissociated from it and be aroused by ideas that stimulate resistance in themselves. But we can voluntarily resist the effect of angry irritation—and even the irritation itself—under the influence of practical considerations, of thoughts of self-respect or the respect of others, or of magnanimous or sympathetic ideals. Indeed, anger that may seem to be uncurbed is generally restrained within limits of resistant volition. When entirely emancipated, it becomes rage. Prudential, or self-regarding ideas, whilst checking angry conduct, will hardly serve to subdue the emotion—may, indeed, exacerbate it. But ideals of magnanimity and sympathy can strike at its roots. Magnanimity restrains it for the sake of overcoming it: sympathy by introducing thoughts that eclipse irritating ideas of antagonism. Influenced by these incentives to self-control, we may be witnesses of a conflict between our physical and spiritual selves. For we may not only refrain from angry manifestations, but honestly endeavour to subdue our temper, continuing the while to tremble with annoyance.

Finally we may remark upon the inhumanity of fear and anger. They are cruel because the one is set upon escape, the other upon retaliation—regardless of consequences. Cruelty may arise from curiosity, the love of excitement, or a desire to manifest superiority. It may be 'politic', in order to procure a practical consequence. This is calculated 'frightfulness'. But the cruelties of war result from anger and nervous apprehension. The former shows itself nakedly in the mutilation of the dead and wounded upon the battlefield. Terror extinguishes every spark of humanity. We may see its effects in the gruesome sacrifices which characterize the more primitive forms of religion, and, nearer home, in the callous indifference of our forefathers to the sufferings of those who were tortured and executed for witchcraft.

CHAPTER XIII

THE AFFECTIONS

Our nervous system manifests its triple character in our affections. They may be distinguished as physical, spiritual, and mental, according as their ultimate origin is in one or other of these planes. In the first class are sexual and maternal love, the attraction of the familiar and the repulsion of the strange. The love of 'attachment' is spiritual: it arises from the association of pleasure with an individual. We owe kindly affection and the glow of sympathy to the unifying faculty of the brain. But this classification is just only if we limit our regard to the *origin* of the different affections. For most of them are, in fact, highly complex conditions. Into sexual and maternal love there enter spiritual and mental elements. Emotional alloys of instinct and spirit are complicated by the action of the brain, and inclinations that are prompted by the brain are infused with emotion.

Sexual love arises as an instinctive craving, and, as such, it is an evanescent animal passion. But it is spiritualized and stabilized by psychic and mental influences. Excellences of appearance and disposition, illuminated by the exhilarating light of love, excite the psychic, or spiritual, feeling of admiration; and to actual excellences the imagination adds enchantments of fancy which excite admiration again. Sexual love at once stimulates admiration, and is stimulated by it; for both are nervous states, and either of them may reinforce the other. Indeed, admiration is actually 'intoxicating'. Community of passion leads the brain to unify the beloved with oneself in ecstatic sympathy, and this unification may be reinforced by similarity of tastes. Consequently love may be focussed by ideas of sympathy in passion, such as brought Benedick and Beatrice together. A passion which includes so many elements would be an effective theme for the novelist, even if it did not possess so enthralling, because so delightful, an interest for readers of all classes. No one will deny that the days of this enchantment were the happiest of his life. And yet in this analysis something fails. Why are particular individuals suddenly drawn together by 'love at first sight'? This can be explained only on the assumption that there is some appealing peculiarity which escapes consciousness,[1] but, nevertheless, instinctively exercises a selective influence.

Why, then, should love, compacted of such strong materials, be inconstant? Partly, it seems, because sexual attraction is followed by a revulsion. Partly because its foundation melts away if admiration is disillusioned, or there is contrariety of tastes. Admiration may have been aroused by one or other of various excellences which seemed to be possessed by the beloved; personal beauty is only one of them. But if the qualities, which were admired, fade, or identity of tastes is discovered to be unreal, love may die away; and the lover may set out to hunt elsewhere for his ideals—the admirable which endures and the sympathy which is real.

Love is less fleeting in woman than in man. In her case revulsion is less threatening to the attraction of sex. Something of the maternal enters into her affection, and this will withstand much discouragement. And, whereas in man love rests upon admiration of the excellent, in woman it owes more to a respect for power which is less likely to be disillusioned by experience. In his character of the breadwinner the least estimable of men is at all events the chief of the household. It is unfashionable to hold that woman regards man as her

[1] Such unconscious influences (possibly olfactory) would account for 'unreasonable' likes and dislikes.

superior, and there is, perhaps, little reason to think that she does so until she is in love. But in that condition she actually glories in her inferiority: it even pains her that she should be taller than the man of her heart: she likes to think of herself as 'little', and to address her lover as 'mon grand'.

On a review of human affections we cannot but conclude that maternal love is the strongest and most durable—and hence the most admirable of all. It may be quite independent of personal credit; it may be free from any suspicion of egotism. A mother may love a crippled child more than one that is admired by her neighbours; her love cannot be effaced by the misconduct—even by the ingratitude—of her son. For the relation of a mother to her child involves identity in its strongest form; as an actual physical fact her baby is 'bone of her bone, flesh of her flesh'. But her love springs in part only from ideas of the brain: it is fundamentally an instinctive impulse to *cherish something*, which is quite independent of any mental process of unification, and profoundly differentiates woman from man. It is proved to her physically that she is destined to support another, and therefore develops more nourishment than suffices for herself. Her nutritive instinct is, however, blind, and must take guidance from the brain. A mother will cherish another's baby—a changeling[1]—if she is unaware of the change. And her maternal feelings may expend themselves in comforting a husband, or in nursing a pet dog.

Another instinctive basis of social feeling is the attraction of the familiar, which, as we have seen, appeals to an elemental susceptibility on the physical plane. It is the ultimate foundation of gregarious life. This is not confined to the higher animals, and is nowhere more elaborate than amongst the insects. We may infer,

[1] Hence the cuckoo—that cocotte amongst birds—can, during the breeding season, find a foster mother for her offspring anywhere, and can use any nest as a foundling hospital.

indeed, that susceptibility to the familiar and the strange is a fundamental trait of life. There is no creature, of however humble an organization, that is not attracted by the one and repelled by the other ; and it is, indeed, evident that a sensibility which actuates the functioning of the brain must be of elemental simplicity. Its effect upon the higher animals, man included, is to incline them towards acquaintances and away from strangers.

Our emotional relations with others are appropriately termed 'affections' and 'attachments', since they draw us to them by linking them to nervous conditions of our own. Hence joy and love are commonly associated in idea : one engenders the other. We become attached to another when we associate[3] him with pleasure. We love those whom we connect with the joy that is caused by helpfulness. Love for one's mother includes feelings of admiration and respect. But, if a mother be unkind, love for her will not develop. Attachments may be formed through *ideas* of benefaction which are suggested to us by others. Heightened by feelings of awe and reverence, such is the love that is inculcated by religion for the Maker and Giver of all good things, and the Redeemer who has freed us from the shame of sin.

Instinctive and spiritual inclinations, however strong, could hardly, however, have brought mankind to form societies larger than the family. For they are only stimulated by those in our immediate vicinity, and could not develop the motives which have aggregated mankind into the communities of tribe, class, and nation. Nor can they generate the intimate affection of sympathy. Social bonds and the sympathetic affections are the fruits of the unifying and differentiating processes of the brain. These of themselves would afford us but a cold appreciation of affinities. But they stimulate psychic feeling, and are warmed by the glow

[3] The shrewder employers of labour know that, to keep workmen well affected, one should pay them oneself.

of emotion, because they involve extensions of our own personality. Altruism is, in fact, an enlargement of egotism which extends our self-love to others.

Unification, as we have seen, may be in kind or in personality. Those with whom we are unified in kind are treated by us with 'kindness'[4]—that is to say, we behave to them as if their *interests* were our own. The traits which by their sameness involve the unifications of 'kinship', are very numerous, and may be ideal, or even imaginative, as well as actually perceived. The most primitive of all identities is that of *family*. This is the bond of brotherhood, which may be extended by the fiction of adoption. It includes the dependents of the family. Hence we may unify our domestic pets with us in kind: the Indian cultivator addresses his plough-bullocks as 'my brothers'. The idea of sameness in family evolves into that of sameness in *race*. This concept can only be formed by inference, and may therefore be erroneous or imaginative. But its influence has been universal. It is the affinity of the *tribe*.

There is sameness in *representation* between those who are subject to the same king or government; and since rulers have commonly obtained their power by conquest, war has been constantly weaving and dissolving ties of nationality. But they may be strengthened by fictitious notions of race relationship. We commonly speak of Englishmen as one in race. Nothing could be farther from the truth. Indeed, judging from the extraordinary differences[5] which may be observed, almost universally, between children of the same family, the races of Europe and America are inextricably intermingled. Sameness of *occupation* unites the members of a profession or trade into a kind-class. It is the basis of Trade Unionism. A railwayman identifies him-

[4] 'Kind', 'kin', and 'king' are derived from the same root (Skeat's *Etymological Dictionary*). 'Gentleness' has a similar derivation: it is the behaviour that we use towards those of our clan.

[5] If a litter of puppies or kittens differ greatly amongst themselves we conclude that they are mongrels.

self with other railwaymen, a miner with other miners; and it is noticeable, especially in Eastern countries,[6] that men of the same occupation tend to congregate in the same quarter of the town, and even to fortify their exclusiveness by hereditary distinctions.[7] The Indian castes are held together partly by tribal or occupational identities, partly by unity of ritual in avoiding or expiating the pollution of impurity. Intermarriage between them would break their unity, and is therefore strictly prohibited. Sameness of *religious belief* is another trait of kinship. It is unnecessary to insist upon its historical importance; but it may be observed that the common observance of a definite and peculiar law has bound the Jews together from time immemorial in despite of differences in race, language, and nationality. Identity in *language* is another unifying trait. So also is identity in *domicile*, that is to say of village, city, or country. This is a comparatively modern evolution, and may be the least moving of affinities. In the East, families of different 'kinds' live alongside one another without contracting the smallest 'local' fellow-feeling. And in town life we are quite indifferent to our neighbours.

In the idea of *nationality* many of these affinities are combined. There is a fictitious notion of identity in race: there is identity in language and in representation—that is to say, in government. In Western Europe the unity of domicile—of country—has become a comprehensive bond of national union, since it offers the great practical (and military) advantages of combining into a whole the heterogeneous classes that inhabit the same area, and of permitting marriage to take place between them. So long as intermarriage is barred, there can be no national unity; and in these circumstances, as in Turkey and India, nationality is merely a plea for

[6] But also in Harley Street.

[7] The rules of the plastering trade, for instance, limit admission, as apprentices, to the sons of plasterers.

self-assertive sentiment. The substitution of domicile for tribe as the basis of kinship appears to have initiated the prosperity of Athens and Rome.

Territorial nationality is, however, in constant risk of being undermined by fraternities of religious belief or professional calling, since, when external danger does not threaten, these evoke a closer spirit of comradeship than subjection to the authority of a single government. It is also threatened by ideas of the racial apartness of communities within the nation—ideas which are commonly attributed to 'race instinct', but are really imaginative fictions. But, since blood-relationship is the strongest of affinities, race-ideals are exceedingly attractive: they may be fortified by more or less fanciful traditions, and even by resuscitating languages which are actually dead. The tie of nationality, when opposed by such an ideal, can hardly maintain itself except by the use of physical force.

A clan, tribe, or nation can be imaginatively personified and expressed by a symbol, such as a tribal totem or a national flag. We feel for the image the reverence of *esprit de corps*, and respect its symbol as a sign of power. A king is, indeed, an emblem of kinship, and nations that have no king lose an imaginative bond of union.

Gender is a *kind*. Men and women are drawn by sex into two different kind-classes,[8] the barriers of which are crossed by sexual and maternal love, and by the closer of other kind affinities. Sexual love rests upon contrariety in sex. Should, then, a woman be unified with a man as belonging to his family, or brotherhood, she loses her sexual attraction. Being identified with him as a sister, she cannot be a wife. Hence in almost all communities there are rules prohibiting marriages between members of the same family. They may go farther and prescribe that marriage must, nevertheless,

[8] Hence, out of the breeding season, so many animals congregate into sex-flocks.

be within the tribe or caste, since the introduction of a husband or wife of a different 'kind' would violate its unity. In fact the rules are 'exogamous' from the family, 'endogamous' from the caste point of view. Our own 'table of affinity' merely insists upon exogamy. Love, as is well known, arises far more strongly between strangers than between those who have been brought up together in close association. Monogamous marriage endeavours to reconcile the two conceptions, by recognizing woman, not only as a wife but as a friend. For a woman may be contrasted in kind, but identified in personality.

It follows from these evolutionary varieties of custom that our social and national institutions have developed through the adoption of different traits as the basis of 'kinship'. An appreciation of them and their effects is the science of social history. One who feels himself identified with a social group is moved by a spirit of active kindness towards its members, since they are one in kind with him: the power of the union excites his admiring enthusiasm, and inspires him with the courage of confidence. Accordingly he is strengthened by his fellowship, and would be weakened by its loss; and the oscillations that give a tidal character to history may, very generally, be attributed to the growth and decay of ideas of kind—to the rise and fall, that is to say, of various kinds of fellowship.

It is to be observed, however, that an idea of identity not only binds men together: it isolates them from other men by suggesting that they are different [9] in kind, or aliens. We may, perhaps, infer that the effect of mere perception—apart from thought—would be to unify all men as of the same kind. For it is noticeable that children will unite with strange children in the intimacy of play—however different they may be in social degree,

[9] This is the difficulty which confronts the State ownership of mines and railways. For Government officials are a class apart, and rarely unify themselves, in kind, with the public.

dress, or even colour—when once the initial effect of strangeness has passed off. But we are continually becoming infected with prejudices against others, as that they are contemptible because they are poor or ill-dressed, or are our inferiors in race, religion, or education—persons who differ from us in kind and should, therefore, be regarded as 'outsiders'. Differentiation may go farther. Aliens become our *contraries*, or antagonists, if injurious ideas are associated with them. By a mental process, similar to that which may occur in perception or thought, the *idea* of them is reversed, as well as the connexion which would link them to ourselves. These isolating differentiations and contrarifications commence with the admonitions of our mothers and nurses as to the persons with whom we should and should not associate. Our prejudices are confirmed by historians and aggravated by our leaders. Indeed, it is hardly too much to say that our affections, and all that depends upon them, are at the mercy of other people's words. A slanderous hint may wreck a friendship and turn into bitterness one of life's sweetest savours.

Let us turn now to unification in personality—that is to say the identification of another with ourselves in spirit. In this case he becomes 'at one' with us, and we 'sympathize' with him—feel as he does. This involves an effort of imagination. For we must personify his psychic traits before we unify them with our own personality. Consequently there can be no true sympathy without imagination. A personality so imagined will be unified with our own if attention is fixed upon its similarities. But these are less obvious than the samenesses that stimulate unification in kind, and, consequently, true sympathy is rarer than kindly feeling. There are, however, points of similarity between every other man—and even every animal—and ourselves, from which we can imaginatively construct a personality that can be assimilated with our own. Assimilation implies union, a concept which in religion has borne fruit of mystic significance.[10]

Unification in personality is *friendship* or comradeship—a more intimate tie than that of kind. The 'pal' is closer than the brother. The emotional effects of sympathy differ markedly from those of kinship. The success of a kinsman evokes our respect, his failure our pity; whereas we are *proud* of the glory of a friend, and share his shame in defeat. We can bear with his faults as we bear with our own. We are merciful when we sympathize. And we have another self to 'love as ourselves'. If we do him a favour we do it to ourselves. Sympathy, then, affords a double pleasure—

> It is twice blessed,
> It blesseth him that gives, and him that takes.

Being imaginative, sympathy can only be kindled by an exciting stimulus. Such, however, is the idea of the Unity of mankind. This is a conception which would spontaneously fire the imagination of few men. The same, however, may be said of our other imaginative ideals. Conceptions of magnanimity, purity, and liberty, for example, must have evolved, in the first instance, from the imaginative aspirations of some particular individuals. But they are scattered broadcast, 'like sparks of fire', amongst mankind by suggestive influence. Ideals are, we well know, evanescent, unless they are endorsed by reason—unless, that is to say, experience shows that they possess practical advantages. But the happiness which sympathy gives and takes is sufficient to impress its practical utility upon the most prosaic of dispositions.

We may, however, contrarify another to ourselves in personality, as being personally opposed to us. In this case we are angered antipathetically. He is not

[10] As that we can identify ourselves with a Divine being by assimilating his substance, and that sin may be vicariously expiated through 'at-one-ment'.

merely an alien antagonist, but an *enemy*. And, since imagination is stimulated by excitement, during a war we are actuated by a personal animosity towards every one of our opponents. As enemies they may be killed without scruple. War is unstained by the guilt of murder. Antipathy can be aroused by ideas, and needs no actual injuries as provocation. It is the force of propagandist suggestion which drives one nation into war with another. Hostility which apprehensions of injury, opposition, or indignity excite in the minds of a few, is spread by suggestion to the mass of a people. So it comes that we draw from imagination the closest and happiest of the ties which can bind man to man, and the widest and most miserable of the breaches that can open between them.

The idea of human unity fixes our attention upon the samenesses that unite others with ourselves. It is pleasurable since it extends our egotistic personality, and can, therefore, be voluntarily pursued as an object. We term it a 'philanthropic ideal'. But, as an ideal, it is peculiar in involving no notion of triumphant resistance, and consequently it does not impress us with power, and evokes no impassioned admiration. It also differs from most ideals in that its promises are borne out by experience, for, if realized, it certainly produces a double happiness. A sympathy which would extend to all mankind is the ideal of all ideals that would bring most happiness in its train.

Yet philosophy has hardly appreciated its advantages, and it was not till the coming of Christ that it became enshrined for admiring desire. From the beginning of speculative thought teachers have been sedulously recommending roads to happiness. In Stoicism and Asceticism they have idealized the pleasure of resistant self-control,[11] in Epicureanism that of the senses.

[11] The Discourses of Epictetus are mainly concerned with voluntary assertive resistance in its various phases. Indeed, life without resistance seemed to him to be merely the periodic filling and emptying of a bag.

Mysticism in its many forms appeals to the awe that is inspired by what is admirable in the strange or powerful: the ideal of Liberty throws open the arena of politics for the winning of dignity. These all make offers of happiness. Priests and professors may call it by another name; but it is the underlying motive of their theories and gives them whatever attraction they possess. But Sympathy surpasses all these ideals, even when regarded in the coldest spirit of analysis. For it offers a double pleasure—to another as well as to oneself.

In the teaching of Christ sympathy was the fountain-head of morality—a spring whose stream could wash away the stains of the Magdalen, could dissolve distinctions of nation and class, and could unite the Greek with the barbarian, the Samaritan with the Jew. Such being the fundamental principle of Christianity, it is remarkable that this religion should not have done more to control the virulence of antipathetic promptings. Man is a plastic creature, easily led by suggestion, and Christ's teachings have been in the hands of influential priesthoods, endowed with ample facilities for spreading them. But who can say that, from the time of Constantine to the present, Christian peoples have shown much trace of Christ's influence in their attitude towards their opponents, in their treatment of conquered nations, or of the poorer classes of their own society?

Christianity has, of course, accomplished much. It has abolished slavery [12] as a *status*. It has established that in theory, and except during war, we should be patient and forgiving in our dealings with others, and that the welfare of others, irrespective of kind, should be our concern. So it has inspired such humanitarian measures as the prohibition of alcohol—the 'spirit' of evil consequences—the enforcement of which by the American people marks the high-water level of philan-

[12] But old ways of thought survive. Witness a recent correspondent of *The Times*, who pointed an account of the moral and material progress of the natives of Rhodesia by observing that 'they were rapidly becoming an important economic asset'.

thropic enthusiasm. We need not wonder that the fruits of Christ's teaching have not been more abundant. In abolishing differentiation in kind, its aim is fundamentally international, and is opposed by all whose interests are served by national exclusiveness. It offers no safeguard against the aggression of those who do not accept its doctrines in literal completeness. Involving no idea of superiority, it makes no strong appeal to admiration and respect; indeed in exalting the humble, it contradicts our natural promptings. Accordingly, Christianity has felt the need of seeking support from mystical sources. Attention has been diverted from Christ's life, with its lessons of practical wisdom, to the mysterious implications of His death. His living emblem—the Good Shepherd—has been supplanted by the dead symbolism of the Cross. His teaching has been obscured by a mass of dogmas and ceremonies which tend rather to strangle sympathy than to foster it. What can be more discordant with the Rule of Love—more calculated to provoke its infringement—than the jealousy which has limited admission to His fold, or the unpitying vengeance of the Old Testament?

CHAPTER XIV

RESPECT, FAITH, AND OBEDIENCE

We have traced the primitive origin of respect to a contractive nervous reaction from the expansion caused by success or power, or by ideas of them. When the power protects us, advantages us, or honours us, the idea of it occasions a further expansive reaction—similar to that of relief. Respect is in this case mingled with confidence, and is emotionalized as *faith*. Accordingly, in its beginnings, faith is an enduring condition of nervous expansion which follows a passing condition of nervous contraction. This conclusion will not seem grotesque if we realize that it is a similar reaction in sequence that gives us fortitude, and may render defeat more profitable to us than victory.

Through nervous rhythmic association the expansive condition of faith may be stimulated without the preceding contraction of respect. If it has become associated with a stimulus, it will be excited directly by the stimulus, whether as an impression or an idea. In this case, through a reversal of the nervous sequence, the contraction of respect will succeed it. So it comes that our respect for one whom we trust seems to follow, not to precede, our faith in him.

Faith is associated with the idea of another when his power protects us from danger, advances our interests, or affords us honour—as is the case if he is identified with us as the chief or representative of our kind. It is aroused immediately we are affected by an impression or idea of him: it is, so to speak, a halo that surrounds him, without which he cannot appear to us. Recognizing him as our protector, champion or chief, we are moved to follow or imitate him, since his interests are our own.

Imitation that is directed by faith is obedience—a truth which suggested the title of 'De Imitatione Christi'.

There are many phases of power. The simplest is that of physical strength: but power also manifests itself in courage, and in the strength of character which resists all inducements to swerve from its purpose. There is strength in self-control. The ascetic has been almost universally reverenced; and the magnanimity which involves self-renunciation attracts respect with magnetic influence. We say that knowledge or wisdom is power. But its power is indirect and impresses us but moderately. Wisdom attracts the multitude only if it pretends to a mysterious influence. The magician has excited more reverence than the surgeon: spiritualism is more thrilling than science. In age there is the power that comes from experience, and consequently we reverence [1] the aged, and extend our feelings to antiquity generally Indeed if we wish to assure another of our respect, we compliment him upon his years: the ordinary title of courtesy—'Sir'—is of course the same as 'Senior'. There is power in numbers, and we are consequently impelled to follow the crowd. Our acceptance of its wishes is in great measure a desire for its approbation. But there is also a more emotional motive. We acknowledge the power of superior numbers, irrespective of other feelings. We admit the authority of a numerical majority, provided that it is identified with us in kind, although we may ridicule its convictions. This is the corner-stone of democracy. Power is a trait of success. It is symbolized by wealth, titles, and decorations. Accordingly we respect them. The mysterious is powerful, for it has unknown potentialities. So also is eloquence, for it may lead the multitude by persuasion.

Ideas of power may be associated with other persons, with imaginative ideals, or with ourselves. We may

[1] A feeling that may be detected in all vertebrate animals that are born in a state of dependence upon their parents.

respect a King, such an ideal as Chivalry, or the personal independence that is involved in Liberty. For they are, all of them, powerful. The King represents controlling authority. Chivalry is strong because it involves resistance. Liberty is the realization of power in ourselves.

Respect for other persons is the basis of all governments, rightly so called, whether civil, military, or religious. It is the mainstay of the family and the school, as well as of the regiment and the convent. Power may be attributed to another by reason of actual experience of his prowess—it may be at our expense. But generally we make the association through suggestion, that is to say, we are taught to regard others as powerful or protective, and respect and trust them, because ideas of them possess traits that are identical with those which have actually stimulated our faith, as a neuro-serial reaction from apprehension. For the vast majority of persons this is the foundation of their loyalty—spiritual and temporal. They have had no personal experience of the power which they revere, and may even be taught to believe that their ruler's authority contains something of the Divine.

When power is concentrated in the hands of one man who, as king or chief, is not merely recognized as protector, but is unified with ourselves as our representative—as the head represents the body—faith becomes the affection of loyalty, and we are proud of him—in spite, it may be, of faults or harshness. Loyalty outlives ill-usage. A schoolboy respects a disciplinarian Head—may even boast of his severity; and amongst Orientals one may still find those who are proud of a ruler because he is tyrannical. In him they feel tyrants themselves.

In these days of committees the energizing effect of a hero has become obscured, and we read without conviction the appreciations of Carlyle. Nevertheless, hero-worship—devotion to a human or divine personality—has generally been the force which has built up the empires and kingdoms that have brought about national, as opposed to tribal, cohesion. And it seems probable that to the commanding influence of despotic authority man owes very much of his progress from savagery to civilization. A despotic government may be abominably cruel and selfish. But it need not necessarily be so: history shows that it may employ its power with good intention to overcome the conservatism of habit and its repugnance to change. The spread of Christianity over Europe, for instance, was impelled very masterfully by royal command. A despotic ruler who means well by his people, and takes good counsel, can change existing customs with far less difficulty than a government which must win its way by persuasion. So careful a student of history as Sir Henry Maine was of opinion that such a drastic innovation as the reform of the Julian calendar would have been impossible under democratic conditions; and in our own days we have seen that so useful an enactment as the Daylight Saving Act could only be carried through under the pressure of war.

Power is delightful, and there is a never-ceasing struggle to obtain it. Kingship may be subverted by an aristocracy which may be as efficient or as inefficient, but will not attract the personal loyalty of those who stand outside its class. A king may be associated with or replaced by a priesthood: indeed an ecclesiastical aristocracy, such as the Church of Rome, may succeed in raising itself above temporal sovereigns, and in establishing a dominion which, apart from its actual authority, is respected because it implies the mystery of a theocracy.

But all these forms of government decay—lose prestige—through a process of gradual but inevitable disillusionment that comes of experience. A conqueror's descendants seldom inherit his forcefulness—not infrequently degrade his attributes. And, this apart, experience weakens despotic authority, as by a process of erosion. The multitude know the monarch only by repute, and may retain their respect for him. But those who are closer to him can judge of his strength for themselves; and if there is any indecision or weakening, they encroach upon his authority by extorting privileges that in effect are fragments of it. So in India the Brahmins undermined the authority of the Rajah by establishing it as a canon that he must act under Brahmin advice; and in Europe the history of the Church, of city corporations, of parliaments, is that of the gradual annexation of monarchical prerogatives. Accordingly the despot gradually declines into the constitutional king, unless, indeed, he have a more tragic ending. For he and his dynasty may be suddenly consumed. If his power has degenerated into a sham—a screen behind which the self-interest of one class preys insolently upon the livelihood of others—conventional respect for him may be shrivelled into nothingness by a blast of revolutionary violence. The community is shipwrecked, and welters confusedly in storm waters, until new centres of power are discovered—new *kiblas* to which men may turn their eyes in confidence. Meanwhile terror rules the community—leaders as well as followers—for in the bankruptcy of faith there is an appalling consciousness of a failure of security. Aristocracies, civil and ecclesiastical, are subject to the same law. The power that is attributed to them is for the most part imaginary, and disillusionment ends in revolution.

But, while it lasts, faith gives extraordinary strength. The power in which it trusts, being imaginary, is unlimited, and one is indifferent to danger if protection be complete. Those who can follow a leader, or an ideal, with uncritical loyalty are untouched by the hesitations which come from fear, or the balancing of consequences. They will face anything, dare anything at his command, since they can apprehend no disquieting possibilities. They are animated with an energy far more resistant than that which springs from pride or considerations of profit and loss. We read with astonishment of the courage of martyrs, of the mental exaltation which enabled delicate girls to face the most appalling and degrading forms of death. And faith gives happiness. There is cheerfulness in the discipline of a convent, of a regiment, if the men 'believe' in their colonel; and military rule is accepted with contentment if the ruler be venerated.

Power may be imaginatively associated with—or attributed to—a material object, such as an amulet, charm, or mascot, and confidence be gained by wearing it. Strongly contrasted with this superstition, but akin to it, is our faith in an imaginative ideal. For an ideal is a thing of power, and we may express our appreciation of its authority by aggrandizing, or 'magnifying', it through conquest or conversion. The fortifying effect of religious ideals has impressed itself on every chapter of history. Those of Muhammad[2] inspired the Arabs with an invincible enthusiasm. The aspirations of Socrates and Epictetus could give strength to meet death smilingly. But ideals fade into impotence unless the hopes that they offer are confirmed by general experience. Their promises are belied by their results, and man falls back to earth from his star. The Crusades were the consequence of an ideal which possessed at the time an immense attraction. They involved great bloodshed, cruelty, and demoralization: their fruits were altogether disappointing, and discredited the ideal which they expressed. The history of man is only in part a record of passionless evolution; in great measure it dramatizes the growth and decadence of idealistic visions—the unceasing contrariety of imagination and experience, the contradiction of illusion by disillusion.

Finally, of power that is associated with ourselves. This gives us self-confidence or independence, the fruit of which is Liberty. One has liberty when he can act

[2] Divine 'Oneness' magnifies power by concentrating it. *Islâm*—resignation to predestined Fate—renders one fearless of *consequences*.

as he likes.[3] He may be influenced by persuasion since this works by the arousing of a like. But he is not subject to command. He is emancipated from the power of others because he has found power in himself. Consequently liberty involves the pleasure of dignity —or, at least, of self-assertion—and this is the attraction upon which its advocates rely. If, for instance, they are addressing an audience of villagers, they exhort them not to salute the parson and the squire: the abolition of the military salute was one of the first-fruits of the revolution in Russia.

A people achieves political liberty when its leaders derive their authority from its likes, as manifested by voting. This gives the electors a feeling of independence and self-respect, which arises irrespective of actual voting. Accordingly, as experience shows, but a small proportion of them would take the trouble to go to the polling-booth, were not special appeals made to their emulation or their private interests. And, for the same reason, an outvoted minority does not lose its sense of power, although its likes have actually been overridden. It has been consulted, and this proves that it has influence. And it acquiesces in defeat all the more easily, since comparatively few persons are really interested in affairs of State. Their intimate concern is with their own affairs, and the interest which democratic politics excite comes in most part from emulation, or from the imaginative pleasure that we derive from a play. If party government is absorbed in a coalition, popular interest wanes. We can then understand why politics affect us so differently to our private affairs. They excite emotion, since they involve ideas of power, whereas in our private business we seek profit by reasoning from experience. 'Les affaires sont les affaires.' Being emotional, our interest in government stimulates

[3] *Ἐλεύθερος* and *liber* are akin to *lubens*, which indicates one who acts upon the impulse of a liking. The original meaning of 'free' is given by Skeats as 'acting at pleasure'.

the imagination, and we accordingly think of politics as 'a game', and do not apply 'business' criticisms to them.

Democracy, in the literal meaning of the word, is of course impossible except in such small communities as the city-states of ancient Greece. Modern democracies are, in fact, oligarchies whose members owe their appointment to the likes of the citizens. Did these cast their votes after reasoning from experience, this system of government would be a progressive evolutionary force. But, as a rule, the likes which influence voting are suggested to the electors, not by experience, but by persuasive arguments which appeal very generally simply to the emulation of party spirit. One persuades another by suggesting likes to him; more effectually still by paying him. Persuasion is the antithesis to command, and accordingly those who aspire to win popular favour must above all things be eloquent. Under a democracy the practical business of the State is controlled by men who have made it their profession to persuade.

Disillusionment haunts democratic, as it does other forms of government. In the first place, self-confidence can never be so sustaining as implicit faith in another, since the power that one feels to be possessed by himself can never be so great as that which he can imagine to be possessed by another. Few men can stand alone. Consequently on occasions of serious danger, losing confidence in self, men search for leaders to whom they may transfer it. It is instructive to observe how, in times of national stress, the most democratic of peoples clings to its leaders, with what virtues it endows them, with what praises it adores them—until there comes the day of disillusionment. It then abandons its hero, but not its trustfulness, and searches for a new chief to whom it may transfer its allegianee.

Habit apart, it is by faith that the poor acquiesce in the enjoyment of accumulated wealth by the rich; and experience shows that, unless a democratic government can gratify their desire for a redistribution of riches, it will be violently assailed by working-class interests. This is the peril of syndicalism, which is the more threatening in that kinship in occupation is more invigorating than a feeling of national unity. And the respect for the power of numbers, upon which democratic authority rests, disappears if the majority of voters are indifferent and apathetic. There comes disillusionment. Numerical superiority appears less powerful than active superiority. In fact a few energetic men represent power more truly than does a multitude of passive electors.

There are other obvious dangers. The key-note of liberty being egotistical, it stimulates self-surrender but feebly, so that even under the threat of great national danger it is inclined to set class-interests above those of the commonwealth, and hopes to continue 'business as usual'. At all times, indeed, vested interests are a drag upon democratic progress, for they will rarely yield to persuasion. Politicians, being dependent upon favour for their position, are liable to rate popularity higher than achievement, and to postpone reform until it is pressed upon them. This is a policy full of danger, since those who have obtained justice by pressure will afterwards use pressure to obtain more than justice. But it is in regard to its dependencies that democracy shows its weakest side. It is unreasonable to suppose that the people of Ireland or India should feel reverence for men because they have been elected to office by English voters. Nothing is more irritating than a disputed claim to superiority: it arouses a jealous spirit of unrest which shows of force will not repress and material concessions [4] will not conciliate. The British democracy has therefore acted wisely in permitting its Anglo-Saxon dependencies to govern themselves; and it appears

[4] The effect of comfort in allaying discontent is over-rated. Is it not said in that book of wise sayings that 'Jeshuron *waxed fat* and kicked'? For centuries the Mohammedans of India repressed all Hindu aspirations by sheer weight of taxation.

inevitable that this policy should be extended to its other dependencies.

Still, with all its defects, democracy offers undeniable advantages. It is less dangerous to mankind than autocracy, because it is less forceful. It cannot repress individual talent with the masterful strength of a reactionary despotism. It may be adventurously aggressive, for its egotism disposes it to be covetous, and if inspired by an enthusing idea, it may be violently militarist. But the commercial democracies of the present day are infinitely less likely to trouble the peace of the world than are autocracies. And, in the second place, the liberty of democracy undoubtedly appears to foster kindly feeling. The notions that conduct should accord with pleasure, and that the instrument of government should be persuasion, breed a spirit of deference which can hardly exist in a despotic atmosphere. Faith, on the other hand, strives to its purpose regardless of consequences. It hardened German patriotism into a cruelty which was insect-like in its indifference to the pain of others, just as, in days gone by, it steeled the hearts of the dons and undergraduates of Oxford against the sufferings of the bishops who were burnt before their eyes. It has a damning record in the history of the Inquisition. If liberty, on the other hand, fosters kindly feeling, those have good reason who sing its praises, and even build upon its extension a dream of universal peace.

CHAPTER XV

EMULATION, INDUSTRY, AND MORALITY

WE come now to phases of psychic resistance which may be classed as emotional because they primitively involve no conscious effort of will. Like courage and curiosity, they are primitively stimulated by conditions of confusion. But, evolved into an associative form, their stimulus becomes an *idea of antagonism* which excites psychic resistance because it has been coupled with resistance in experience. The idea of resistance *to* us stimulates associatively resistance *within* us. When the antagonism is between two personalities—our own and our idea of another's—our *emulation* is aroused. There is antagonism between us and our material environment, in that it affects us against our wishes and cannot be changed without an effort. An idea of this antagonism stimulates us to the resistance of *industry*. And, thirdly, there is a standing antagonism between our physical and our psychic natures. The automatic sequences of stimulus and response, and the dull routine of habit, are incompatible with the mercurial vivacity of our 'spirits'. This incompatibility causes them to be sharply differentiated by the brain—an antithesis which is heightened by the imagination. Resistance to physical promptings is the ultimate source of *morality*. Accordingly, emulation, industry, and morality originate as subconscious phases of assertive volition. But they evolve into motives that are consciously voluntary, since the pride and shame which follow success and failure stimulate us to the selective volition of choice.

We can infer from observation that *emulation* is primitively a phase of subconscious resistance. How inevitably does horse run against horse, dog against dog, one thrush or canary sing against another! We can detect in ourselves the action of this subconscious motive: it is true that we may voluntarily set ourselves to excel others, but it is also true that we are affected by the spirit of rivalry quite unthinkingly. The hunting spirit is fundamentally emulative: we 'pit ourselves' against the animals we pursue, although this feeling of rivalry only becomes evident when we subject ourselves to analysis. The spirit of emulation enters into the enthusiasm of war. But rivalry does not necessarily involve anger, although this emotion may arise from the shame that follows a rival's success.

Industry or perseverance sets itself against the rivalry of inanimate things. When we are engaged upon a task it appears to strive against us, and consequently, if we succeed, we have vanquished it, and experience the revulsion of pride. But, fundamentally, our industry is actuated, not by conscious anticipations of triumph, but by an automatic response to an idea of antagonism. Ordinarily, we have no thought of pride when we set ourselves to accomplish a task.

Nevertheless, success in subconscious emulation and industry is followed by a conscious feeling of superiority or pride. And since we are in continual contrast with our fellows and our environment, and success can be won in the most trivial matters, our normal state is one of self-complacency. We can maintain it in successfully discharging the most ordinary duties of everyday life. This feeling, being the result of rising *superior* to our environment, is the sharpest distinction between ourselves and environal influences. In fact it is oneself. Our self-complacency takes various forms, and bears various names, according to the idea with which self is continuously associated. It is *conceit* if this idea is simply one of success: *vanity* if it is that of the respect of self or of others. A *self-assertive* habit of mind proceeds from the association with self of the complex idea of dignity, in which power carries an implication of respect. If this succession is reversed so that the associated idea is that of honour, it renders one *arrogant* —that is to say, exacting of respect.

When we are *voluntarily* emulative or industrious we pursue the consequence of our efforts as the pleasure of success: we are stimulated by the expectation of winning superiority and a feeling of pride. This is ambition. Similar ideas, we have seen, excite voluntary courage and the curiosity of explorative thought and adventure. Ambition may be either self-regarding or ostentatious. In the former case it urges us to excel for the pleasure of excelling, either by emulating others or by industrious perseverance. It is the spirit of active self-assertion. We dislike it in others because it excites our jealousy. But our debt to it is, nevertheless, incalculable. For it affords the reformer, the inventor, the artist, and the explorer the energy which they throw into their work.

Ostentatious emulation may be equally practical and useful. But it is incited, not by visions of accomplishment in themselves, but by thoughts of the effect that accomplishment will produce in attracting the respect of others. It is, then, *arrogant*. We have seen that respect and success are nervous conditions so closely connected that an idea of one recalls a feeling of the other. Hence we feel proud if we are respected, admired, or popular, dignified if we impress our companions or the public, so that the triumph of success comes to us indirectly through the approbation of others. Self-regarding, or sincere, and ostentatious ambitions can exist together, and we may often be puzzled to discover which is our real motive.

Moreover, since the pride of success can be won by impressing others, ambition may be simply pretentious —that is to say, respect may be pursued without attempting to merit it, in the assurance that dignity will come with it hand in hand. All that is necessary is to cultivate esteem or achieve popularity. For others can judge us by appearances only, and may, then, be easily misled by pretences of successful achievement or superiority.

Let us now turn to *morality.* Actions that are moral differ from those that are natural because they involve resistance, and are followed by a feeling of pride or self-complacency. 'Virtue' implies effort. It is difficult to realize that virtue may be automatic. Yet we can feel that our moral actions are not solely motived by thoughts of their consequences—that we are impelled to be moral irrespective of advantages. And if we trace moral rules to their ultimate source, in the archaic superstitions that are studied by Anthropology, we shall find much to assure us that they have grown out of irrational resistances. We may seek virtue voluntarily, in order to win the pride of success or to avoid the shame of failure. But we are impelled towards morality by ourselves—by an incompatibility between the physical and the psychic which is dramatized by the imaginative personification of the two as opponents. This antithesis enters into a large number of religious ideas. Indeed the conception of diabolic influence has arisen from the insistent antagonism (or *contrariety*) of our physical promptings and our psychic aspirations. The devil is within us, not outside us: so also is the kingdom of heaven.

The evolution of *voluntary* morality has been so complicated a process as to defy analysis unless we can submit ourselves to the humiliating idea that the etherialized conceptions of civilized man have their roots in such superstitions as survive in the Indian caste system, and in the social practices of uncultured tribes. From this point of view it seems evident that the pride of successful resistance to physical impressions, promptings, and ideas is primitively figured as the dignity of 'purity', because it represents the overmastering of sensuous influences which, if unsuccessfully resisted, involve the contamination of shame. Consequently dignity is associated with the energy that is within man and pollution with the influence of the senses, or of the physical body, and primitive imagination employed itself in listing and grading the various activities or influences which pollute if unresisted, and the remedies —that is to say, the resistances—by which pollution might be removed and the dignity of purity be re-asserted. This is the magical state of morality—prohibitive, not suggestive, typified by the *taboo.* It is a petrifaction that has been formed in the springs of assertive volition. And since imagination uses analogical similarities for extending its inferences, ideas of the things which pollute and of the remedies for pollution, may be multiplied indefinitely, so that magical observances and rites become complicated out of all resemblance to the ideas from which they originally sprang.

Purity is at its height in the dignity of asceticism—the systematic resistance of all physical temptation, which gives a pride that can compensate for the loss of all sensuous pleasures. In a less aggressive form it becomes self-denial—exercising itself in the performance of acts of self-sacrifice which may conduce very greatly to the happiness of others, but win their own gratifying reward. That this is so is often appreciated, and acts of self-denial may be resented by those for whose benefit they are performed as puritanical endeavours to 'save one's own soul'. But the world would be a poorer place without them. Muscular movements and utterances are physical, and we can therefore resist the manifestation of emotional excitement, although it may retain its influence upon us. So we may resist the expression of pride or anger: words may be repressed in a dignified silence. Indeed a sense of dignity may be expressed by humility.

Morality outgrows its magical swathing-clothes and becomes *practical* when it is directed by an appreciation of the material consequences of conduct. Adultery is condemned, not because it involves pollution, but because it is harmful to society. Expiation evolves into punishment, with the function of deterring men from committing immoral acts in the future, not of cleansing them from the consequences of immoral deeds in the past. But the two conceptions are often confused in modern thought. We speak of a convicted criminal as 'expiating his offence by imprisonment', and we figure sin as 'polluting'.

There is a further refinement. Morality becomes *ideal* when its rules are associated with ideas which render them attractive, such as those of magnanimity, or resistance for its own sake, of sympathy with our fellows or of obedience to a revered authority. But without undue cynicism we may conclude that the strongest of these idealizing conceptions is that of our own dignity. So fortified, the most arbitrary rules in regard to dress and conduct become irresistibly attractive. Although, however, moral observances may have a utilitarian purpose, or may be endorsed by a like or dislike, we feel, nevertheless, that something more is involved—that they have a deeper foundation than a regard for profit or loss, pleasure or pain. And our feelings do not deceive us. For behind these considerations there lies the fact that moral resistance is primitively subconscious, and that we have something within us that resists,[1] irrespective of our consciousness of consequences.

The moral rules of a particular class of persons or a period rapidly become *conventional*, that is to say, become liked because they are familiar or habitual. It is sometimes asserted that morality is wholly conventional—that it is simply a reasoned expedient for the protection of vested interests, for the maintenance of the orderliness which is essential for the accumulation of riches. Not so. The impulse which it represents springs from the inmost recesses of human nature. But we must admit that it represents this impulse with extraordinary artificiality. For, since pride can be won

[1] 'Life is more of a Wrestling-match than a Dance' (Marcus Aurelius, *Med.* vii. 61).

by resistance in any form, the rules of morality may enforce any kind[1] of resistance. Their observance is obligatory, because the alternative of non-observance involves the painful consequence of shame: we are consequently tied to them: they represent our 'duty' however fantastic they may be. The rules which enforce modesty and decency, for instance, are of such remarkable diversity that they can express no inherent propensities. A generation ago modesty forbad a woman to show her ankles. Chastity admits of polygamy in some countries, and of polyandry in others: indeed, some peoples have found it consistent with respectability to discard formal marriage altogether, giving women, instead of men, the control of property. There are astonishing inconsistencies in our definitions of crime. A man may not steal a morsel of food, but by speculating or profiteering he may deprive others of hard-won earnings. Penal servitude is the fate of one who revenges in private life a deadly injury: glory is the meed of another who, in time of war, kills numbers who have done him no harm whatever.

There is, then, the shadow of an excuse for anarchist reformers who denounce moral and social conventions as being simply expedients which have occurred to the richer classes for their own advantage. But the idea of dispensing with them altogether is absurd. They realize connexions between causes and consequences which, although often established incorrectly, or in selfish interests, are at least an attempt to provide against the future. Herein is the greatest of the differences between human manners and those of the brutes. It is convention that protects us from the tyranny of brute force. Its rules are artificial: so are our clothes, but no one suggests that we should go naked. Many of our conventions are stupid and unjust: others have contributed

[1] Montaigne (*Essays*, i. 22) amused himself in compiling a list of anomalous and inconsistent eccentricities in morality. Some of his instances are exaggerated. But he proves his point—that 'good form', in great as in small matters, is determined by custom.

immensely to man's happiness. Such is the rule that we should 'play the game', that we should desire, not only to win, but to win fairly. Our hopes for the future lie, not in condemning convention, but in reforming it by modifying such of its precepts as are out of accord with the sympathy which we can feel for others if we identify ourselves with them.

It is noticeable that the motives that urge us to observe the rules of morality are threefold, and correspond exactly with the springs of ambition. For moral conduct that is willed may be conscientious, ostentatious, or pretentious. In the first case we seek the pleasure of triumph directly, and are influenced by Conscience—the image of our psychic selves. Ostentatious morality, on the other hand, pursues this pleasure through the approbation of others. This will be won by good conduct, and will be accompanied by a feeling of successful effort. A man's motive may be, not to win approval, but to escape disapprobation. This will give his thoughts a pessimistic tinge, and he will be characterized, not by a sanguine temperament and a smiling expression, but by the acidity of disposition and the gloominess of feature which we associate with puritanism. But, however this may be, ostentatious or pharisaical morality achieves practical results: it accomplishes virtue.

Pretentious, or affected, morality may, on the other hand, bear no practical fruit whatever. It rests upon the maintenance of appearances. For, by impressing others through the mere pretence of virtue, one may gain for himself such a feeling of pride as would accompany resistant effort. He is assisted by the conventionality of moral rules; for, when conduct is liked because it has become familiar, nothing more than imitation of familiar appearances is needed in order to gain popular respect. So one who takes a cynical view of religion may obtain a glow of pride by regular attendance at church. Indeed his Sunday clothes will suffice to dignify him, because, being conventional, they impress his neighbours with an idea of moral tone.

That moral pride may be won by sheer pretence encourages the hypocrisy which grows, like a fungus, upon our virtuous sensibilities. And, since pretentious morality addresses itself to the judgement of others, it will not maintain itself when the eyes of others are removed. Self-regarding, or sincere, morality, on the other hand, is reliable, for conscience is always present. The conscientious man is never alone: he is always under the watch of a critical companion, and can never enjoy the freedom of solitude.

Self-respect, and the respect of others, may then lead us into very different courses. Yet one is the origin of the other. For we only seek the good opinion of those who are *unified with us in kind.* We court the approbation of those of our own class, and are indifferent to the views of persons who are much above or below us in the social scale. An Englishman in the tropics troubles himself not at all as to the feelings with which the natives regard his conduct. He does not deny himself the pleasures of drinking or dancing because in their opinion these distractions are immoral. On the other hand he is exceedingly sensitive to the opinion of his class as to what is or is not 'good form'.

CHAPTER XVI

THE EVOLUTION OF PLEASURES AND DISPLEASURES

To some Determinist philosophers it has appeared that pleasure and displeasure are unessential adjuncts to living activity—mere *epiphenomena*, which may be likened to the sparks that are thrown off by an electrical machine, but do not contribute to its effectiveness. It would, however, be extraordinarily anomalous were the feelings that give life its conscious value evolved for no definite purpose. And it has already become clear that, so far from being useless, they are the instruments of a process of *conscious* evolution that has been grafted upon that which is unconscious. For, with the refinement of pleasures and displeasures that marks the progress of civilization, methods of pursuing the one and of avoiding the other, and of manifesting them, naturally increase in complexity.

It is true that the pleasure or displeasure given by a stimulus that touches us instinctively is simply the conscious reflection of an instinctive condition. The stimulus primarily affects us unconsciously, and its agreeableness or disagreeableness is merely an ancillary accompaniment. But this accompaniment introduces the possibility of an evolutionary development. For a pleasure that has been experienced in the past may be expected in the future. It then becomes an incentive to its conscious pursuit,[1] and the methods of pursuit, being determined by pleasurable or displeasurable consequences, will leave the beaten track of instinct and evolve into conscious courses of conduct. And pleasure

[1] One does not eat solely for the pleasure of eating. But from this pleasure is evolved the conscious pursuit of delicacies of food.

being our motive, our aims will grow in complexity as pleasures are elaborated. They obviously increase in number with advancing culture. Contrast the pleasures of civilization with those of savagery. They are as a thousand to one; and the variety of methods of conduct is in a like proportion. 'Progress' is the increasing complexity of pleasures and of methods of securing them. We are continually refining simple instinctive likes and dislikes, and so acquiring fresh tastes that are non-instinctive, and affect us merely through the pleasurable or displeasurable character of the ideas that represent them—tastes in food, colour, and sound which differ in an extraordinary fashion from nation to nation, and have varied continually during the past. The Chinese and Europeans are mutually astonished by each other's tastes in food, in dress, and in art. We cannot understand how the Romans could have carefully reared for the table curiosities of marine life which to us are repellent. The study of the evolutionary refinement of pleasures and displeasures is, then, a necessary preliminary to that of voluntary behaviour.

All pleasures and displeasures involve the intervention of the brain, since it is this which renders them *conscious*, by combining motor excitement with sensory excitement. When the motor excitement is simply the effect of sensory impressions upon the brain—that which is transmitted from one idea to another in the process of thought—it gives pleasure of the subdued type which we describe as 'interesting'. It becomes aggressively pleasurable or displeasurable when it is sympathetically controlled or intensified by the vibrations of motor excitement on the psychic plane. Impressions, whether of outside things or of our own organic and nervous conditions, energetically please or displease us when the cerebral motor excitement that produces consciousness comes under psychic influence. It is through ourselves that we are affected acutely by our environment.

Pleasure and displeasure, in themselves, are nervous conditions of expansive or contractive vibration in varying degrees of intensity. They become particularized as definite pleasures or displeasures by being associated with the particular stimuli which arouse them and their physical sensory effects; and since stimuli are of very different kinds and may affect us perceptively, memorially or imaginatively, or partly in one of these fashions and partly in another, the classification of pleasures and displeasures is no easy task. But, having regard to the nervous conditions which accompany them, we may classify them *in kind*, as physical, emotional, spiritual, and intellectual. From physical pleasures and displeasures are evolved those which we distinguish as aesthetic, artistic, romantic, idealistic, egotistic, and economic. 'Spiritual' pleasures and displeasures are developments of psychic revulsion. Emotional and intellectual pleasures are explained by these names. Pleasures and displeasures may further be distinguished *in phase*, according as their stimuli present themselves in actuality, or are memorial or imaginative. But imagination may be an element of kind as well as a cause of phase: idealistic pleasures, for instance, are imaginative.

First, then, of physical pleasures and displeasures. In their simplest forms they are those of the senses—the consequences of sensory impressions that are instinctively agreeable or disagreeable, such as those of pleasurable or painful touches, internal or external, and of crude impressions of smell, taste, sight, sound, and rhythm. They also include the satisfaction or disappointment of the appetites; comfort, and discomfort. There are other physical stimuli which influence the psychic plane in a peculiar fashion, with consequences that overshadow their physical effect. Such are alcohol, drugs, and music. The two former appear to stimulate the nerves after the fashion of the *hormones* which are secreted by certain peculiar glands. The *rhythm* of music appeals to a physical susceptibility. But its *notes* as they rise and fall, alone or in combination, seem to touch the psychic plane directly. We must remember that they are actually atmospheric vibrations of various degrees of rapidity:[2] they are rhythmic in themselves apart from the rhythm of the time in which they succeed one another. If psychic excitement is also vibratory, as we have inferred, there is a similarity—a sameness of trait—between emotional conditions and the nervous effect of concords and discords in music. Harmony is concord in coincidence: melody concord in sequence, in that each note is rhythmically familiar with its predecessor. Discord in coincidence, and unfamiliarity in sequence, produce upon us the effect of the strange. And the resolution of a discord gives the pleasing effect of a nervous revulsion. The effect upon us of a major and a minor key has something in common with exhilaration and depression. Music has a further resemblance to emotion: it is characterized by varying degrees of pitch, intensity, and rapidity. It would be an interesting exercise to develop this comparison between emotional conditions and those stimulated by music. But it would be out of place in this book, and it must suffice to infer that music affects us unconsciously as well as consciously. It may touch us during sleep: nurses are aware that a lullaby soothes a sleeping infant. But its *beauty* can, of course, be only consciously appreciated—that is to say, ideas of its melodies and harmonies must be formed.

These natural pleasures evolve into extreme complexity by the admixture of 'spiritual' and imaginative elements, and by the influence of appetitive expectations. The differentiated tastes called 'aesthetic' owe their attractiveness to the association of spiritual ideas of power or excellence with particular traits of sensory impressions or feelings, which are appreciated through

[2] Since vibrations are numerical, Pythagoras was not altogether fanciful in thinking that music and mathematics were connected.

a growing refinement of sensory susceptibility. This development comes about through the repetitions of practice, by which a man learns to appreciate, for instance, not merely the difference between port and sherry, but between various brands of sherry. Sensory appreciation is sharpened when a sense is frequently exercised by being subjected to slight differences of the same kind of impression. A professional tea-taster refines his sensibility by continually tasting various grades of tea. It has been proved [3] that tactile susceptibility gains delicacy in one who is subjected to experiments testing his susceptibility, since these give him practice in feeling touches, and discrimination becomes more acute. This is in accord with the experience of everyday life. The more often we see good pictures, or listen to good music, the greater is our enjoyment of their beauties. The traits which we appreciate may be instinctively agreeable. But for the most part the pleasures which they yield are derived from the association with them of ideas of *excellence*.[4] This implication illustrates very strikingly the chain of associations through which pleasure and displeasure are communicated to impressions which in themselves are colourless.

Excellence is a trait of success, pleasurable because it associatively excites a psychic expansion representing the revulsion of pride that follows success. Therefore qualities which imply excellence are pleasurable and excite admiration—that is to say, a like accompanied by a psychic expansion. The attraction of excellence is what we mean by 'beauty'. Since, however, success depends upon the standards that are aimed at, and these may be conventional, or decreed by fashion, aesthetic tastes differ extraordinarily, and have varied as one type of civilization has developed into another. For fashion is imaginative, and changes as new fancies arise and are spread by the influence of imitation, persuasive sugges-

[3] Schafer's *Text-book of Physiology*, ii. 936.

[4] As is recognized in the phrase 'I can find *nothing in it* to admire'.

tion, or the prestige of those who adopt them. *Power* is also a trait of success, and we are aesthetically impressed by a trait which implies it. Rapid movement without fatigue is attractive because it is powerful. There is a further complication. The trait of *peculiarity* touches us aesthetically because it implies dignity. For our own distinctive peculiarity is the egotism of our dignity, and peculiarity is, therefore, dignified and arouses our respect. We can hardly refuse respect to those who are 'exclusive'. And since an idea of power is associated with age by the dependency of childhood, we respect the aged and find an aesthetic pleasure in the *antique*.

It appears, then, that our aesthetic tastes owe their pleasure to associated ideas that are derived from our psychic experiences, or feelings. These kindle admiration, which, it must be realized, is attended with psychic excitement that is as actual as that which is involved in alcoholic intoxication. It is a nervous state of expansive vibration. Anything that we admire gives us a real pleasure, even although our admiration has been excited by the most fanciful or conventional of excellencies.

It follows that we are continually adding to our pleasures, since we refine the tastes which we frequently exercise, and their refinements, being peculiar, are pleasurable. Water-drinkers appreciate differences in the quality of water that are not detected by those who prefer stronger drink. By the exercise of our senses new shades of pleasure in colour, form, and sound are evolved which have no foundation in instinct. It is literally true that

> with every morrow are we weaving
> A flowery band to bind us to the earth.

When a differentiated taste has become familiar, it acquires a new zest, for familiarity is attractive in itself. Familiarity depends upon memorial association, and if this dissolves, susceptibility to the pleasure of a taste fades away. Those who have become strangers to luxuries find on returning to them that they have lost their savour; and one may, beyond doubt, lose artistic sensibility [5] by failing to exercise it.

Few aesthetic tastes have evolved from the pleasures of touch, the most archaic of our senses; the pleasure of exercise for its own sake is an instance of the exceptional. But out of scents, tastes, colours, forms, and sounds, shades of pleasure have been refined in almost infinite variety. The production of their stimuli is a form of Art. We apply this term to language as well as to painting, music, and cookery. For we have tastes in literary style as in other sensations. Pleasure in the *beauty* of music is an artistic taste. But, since its excellence depends upon our standards, and these are largely conventional, fashions in music change with time, and differ from one people to another. A European must alter his standards of taste before he can appreciate oriental music.

There is excellence in feeling, action, and speech, as well as in sensation. The beauties of the emotions are naturally inherent in them, and not conventionally attached to them, and are therefore unaffected by fashion. All the world over, courage and pity are admired and magnanimity honoured. We admire dexterity in muscular movement, eloquence in speech, and the graces of literary style because they imply power—and this is a trait of success. But they typify it conventionally: different peoples admire different dexterities and different forms of speech. It may be observed that the aesthetic pleasure given by eloquence has momentous consequences; for it induces us to choose our leaders from amongst those who excel in speech-making.

Aesthetic pleasures evolve into the *artistic*, when excellencies are personified and decorated by imagination and are figured by emblems. In the artist these

[5] Darwin, engrossed in his scientific studies, was surprised to find that his taste for music had deserted him.

re *creative* : in the admirers of his works they ...ppreciative. Painting, sculpture, and architecture emblematize excellencies of superiority or peculiarity. It is of the essence of an emblem that it should be tangible. But gestures involve tactile impressions, and are used emblematically in dramatic art. And, since speech also involves feelings of touch, words are emblems (or symbols), and the creations of imaginative Art can be expressed by language, or in the written composition of poetry and fiction.

When methods of conduct are imaginatively decorated with traits of excellence or peculiarity, they afford the pleasure of the *romantic*. There is romance in a picnic, or in a fancy-dress ball, in riding upon the top of a taxi-cab, in all phases of conduct which are called 'sporting' or playful. Children are passionately romantic because they are highly imaginative, and welcome any method of accomplishing an object but the 'reasonable'. For a similar reason the exhilaration of love tinges courtship with romance—which is conserved in the courtesy with which women generally are treated.

Idealistic pleasures are afforded by traits which excite strong admiration and are imaginatively figured as tangible individuals. Our ideals are personified qualities that have been individualized as continuities. They can, therefore, be easily expressed in painting and sculpture. The traits which they represent may be aesthetic, emotional, or spiritual. We may be thrilled by ideals of Beauty. Love, Courage, and Sympathy represent emotional traits that involve great pleasure and are therefore inspiring. But the strongest of our ideals are spiritual. Most of them individualize traits of the success that becomes psychic as triumph. Thus Courage represents the strength of resistance to fear: Dignity, power coupled with respect: Honour, respect coupled with power or excellence: Liberty, self-power: Duty, the strength of resistant morality: Truth, the power that is manifested by successful efforts of curiosity, or that is involved in frank avow... and therefore excellent and admirable, b... us from the indignant surprise that is ca... reward of vice or the punishment of virtue. It is ... for justice that renders us *grateful*. Magnanim... represents resistance for its own sake. This precedes the pride of success in the chain of causes and consequences, and is, therefore, the most transcendental of ideals. It is the keynote of the self-renunciation which displays itself in the self-control of the higher asceticism, in charity, generosity, patience, and in the forgiveness of injuries—and will not desert a friend because he is 'down' or a cause because it is losing.

When aesthetic traits are associated with oneself, the pleasure or displeasure which they afford is *egotistic* or self-conscious. This association is imaginative, since it involves the subdivision of self into three personalities, one of which—the reflective—appreciates its physical and psychic counterparts. So one may admire or despise himself according as he manifests strength or weakness, excellence or inferiority, and may experience a self-conscious pleasure or displeasure, which to the imaginative is extraordinarily poignant.

Pleasures and displeasures may be those of *future* expectations—that is to say, of expectations of expectations. They are stimulated by things which are of advantage in building up a reserve against future expectations, and we may, accordingly, term them *economic*. The attractiveness of a thing on this account is distinct from its pleasurableness in *present* expectation: it is its value in the eyes of the *prudent* man, as opposed to him who 'takes no thought for the morrow'. Material civilization is based upon economic pleasures, for it is these that incite man to make or procure the useful, in view of exchanging it, and to accumulate stocks. Future, rather than present, expectations of comfort urge us to acquire possessions: hope is the 'magic of property which turns earth into gold'. But money is the object have seen, is a concept in which an s linked to one of respect. The pride timulates is one of the most poignant of s: in youth it may yield to love, but as age vances it becomes life's greatest pleasure. The appreciation of it is egotism. Fortunately for man, he ordinarily has a good conceit of himself. There is no one who in his heart of hearts does not believe that, in some respect or other, he excels his fellows—if only in his misfortunes, or in his capacity for absorbing drink—and if we try to 'catch ourselves' in thought we may be surprised to find how often our own credit is the subject of our reflections. The pleasure of patriotism is egotistical: we pride ourselves upon the country with which we are unified, and cannot bear that its dignity should be lessened. But there are moments of disillusionment, when thoughts of inferiority oppress us; and, consequently, reassuring compliments please us greatly. The consciousness of enduring shame is the most miserable of all conditions, and may even drive a man to rise superior to himself in the supreme resistance of suicide.

A sense of superiority may be given us by emblems or symbols. Titles and decorations afford this pleasure, and dress is appreciated more for its symbolic significance than for its comfort. As the distinction which it gives is artificially associated with it, fashions in dress may be quite as arbitrary as fashions in decorations. The pleasure that is derived from imposing houses, furniture, and equipages is in the main that of pride. Money is a symbol of superiority, and is probably valued more on this account than as an instrument of luxury. But each nation gives forms of its own to the coins that represent wealth and are, therefore, the symbols of a symbol. Superiority may be symbolized by a stately deportment, which consequently gives pleasure to him who affects it.

Any object or form of activity may become a symbol—or 'source'—of pride, and hence things and actions which primarily are merely instrumental—that is to say, are of value through a relationship evolving into stimuli that are attractive Language, from being simply an instrument o sion, becomes a science which is studied for its own s Education is transformed from a means into an end, and the art of government becomes an elaborate complexity of procedure, which affords dignified pleasure to the officials who are charged with it, and, in their hands, incessantly evolves new tastes in red tape.

A sense of pride or shame may also come to us indirectly through the esteem or disesteem in which we are held by others. For, since a feeling of respect or esteem is the neural consequence of one of power or excellence and the sequence is reversible, an idea of another's respect or esteem recalls a feeling of power or excellence. 'Honour' is an idea of esteem coupled with one of power or excellence. Accordingly self-esteem and popularity are pleasurable, self-contempt and unpopularity displeasurable. We may achieve this indirect pride symbolically. Dress wins esteem if it is conventional, and, since standards of convention vary, fashions in dress are extraordinarily diverse. The combination of a frock coat and straw hat, which shocks an Englishman, excites no comment across the Channel. But, since superiority is won by distinction, fashions in dress are swayed by two contrary tendencies—towards the peculiar and towards the customary. The symbols may be manners as well as things. These exhibit curious vagaries in symbolic association. The Chinese actually thought that they added to a woman's beauty by mutilating her feet, because small feet had become a mark of social distinction. So have developed the curiously different customs for the disposal of the dead. We dislike the idea that the bodies of those whom we have loved should be disposed of otherwise than by burial: the Hindus prefer cremation: the Parsis, a sensitive and intelligent race, can suffer without dislike that their dead should be torn and devoured by vultures. Each of these methods has respectability: there is no family so to be tempted into extravagance by the an ostentatious funeral.

eas of superiority or inferiority in others afford us the pleasure or displeasure of pride or shame *if we sympathize with them*. Otherwise they affect us as associative stimuli of re-active emotion, our respect or pity being excited if those with whom they are associated are of our own kind, our jealousy or scorn if they are differentiated from us.

The psychic pleasures which we term *amusements* are very peculiar manifestations of psychic revulsions that are akin to those of joy and sorrow. Their course is hidden from us, because they result from a subconscious mental process. Their stimulus is something which causes surprise, alarm, or respect that is immediately dispelled, so that there ensues such a psychic revulsion as occurs when, having been causelessly frightened, we 'smile at our fears'. A shock is succeeded by reassurement when expectations are affected by the fluctuations of 'luck'; when the fearful in imagination does not possess the trait of reality; and when a connexion of ideas, which is strange in association, is familiar in trait. From these contradictions there results the amusement of the fortuitous, the mysterious, and the ludicrous. Since the revulsion is occasioned by an idea-trait, it is generally so rapid as to escape analysis. And psychic expansion being the normal condition, the expansive rebound is far stronger than the contraction which preceded, and, if unchecked, frees itself in smiles or laughter. When, however, revulsion takes the form, not of a single sudden rebound, but of a series of rapid oscillations, there is no such need of a safety-valve. In playing cards, or in reading a ghost-story, we are amused; but we do not break into laughter, unless a ridiculous incident occurs.

The amusement of the fortuitous is caused by the 'lucky' ups and downs which occur in actual life and are invited by us in the playing of games. In all games that amuse there must be an element of luck. This of itself may not be amusing: we experience a revulsion of joy, not of amusement, if we accidentally light upon a sovereign. But if good luck interrupts a state of nervous contraction, caused by the thoughts of possible loss, it occasions the rebound which is the essence of amusement. These are the conditions of gambling. To its amusement is added the pride and profit of winning. We need not marvel that gambling and betting are the habitual pleasures of multitudes of mankind.

A similar revulsion gives pleasure to the mysterious. Its incidents are alarming in imagination, but not in actuality: our apprehensions are allayed as they are raised, and afford a succession of pleasing reassurements. There is nothing that amuses in the really mysterious: in awaking at night, for instance, and hearing the soft pad of footsteps about the room. But we are amused by the relation of this experience, because we are not actually affected by it; and each contractive thrill, as it occurs, is a spring which sends up an abounding sense of enjoyable security.

But the most remarkable of all amusing revulsions is that which affords us the pleasure of the ludicrous, comic, or humorous. Various theories have been advanced to explain its source, one of the best known of them[7] maintaining that ridicule is a weapon that has been developed by evolution in order to check eccentricity that would be harmful to mankind. But a leading clue to its nervous origin is given by the smile of relief which follows an alarm that proves to be groundless. It manifests the pleasure of revulsion. Now in the comic, or humorous, there is always some trait which surprises—which 'takes one aback' by being unfamiliar in its connexion or context—that is to say, by being inconsequential,

[7] That of M. Henri Bergson. But it is difficult to reconcile it with the fact that we may be amused by manners or language without expressing our feelings in laughter, and that we may laugh at a thing with no sense of disparagement.

unexpected or 'paradoxical'. The shock may be so slight as to be subconscious; but it causes a psychic contraction nevertheless. Its effect is dispelled by a sameness in character or relationship that familiarizes the strange trait with its connexion, by harmonizing the two with a sequence of perceptive or reflective experience. Through 'second thoughts', intelligence comes to the assistance of memory. An expansive reaction follows —of far greater energy than the contraction of surprise which preceded it—and this liberates itself in smiles or laughter. The process is so rapid that we cannot always detect the surprise that is caused by the first thought and dissipated by the second. But, with those who are slow in seeing a joke, it may occupy an appreciable interval.

Thus we smile at children with 'grown-up' manners —at adults who behave like children—because their manners, although strange, are nevertheless *human.* We are amused at inconsequential connexions between thought and language, or between the ideas which commence and end a sentence, because, after all, they are *possible.* Our amusement is vastly increased if the connexion which is made by intelligence has the effect of *bathos*—that is to say, if it humiliates a situation the dignity of which has caused a psychic contraction of respect. For an expansive reaction follows, which reinforces that caused by the familiarization of the unexpected. So we are greatly amused if stateliness is discomfited, and even by the confusion occasioned in a railway carriage by a sudden jerk of the train.

Jokes are suggestions of ludicrous situations. Being expressed in words, sameness of sound between different words becomes an instrument for familiarizing the strange with its context, and we are amused by a pun and by a *double entente.* The point of a joke is enhanced by a bathos,[8] and this gains in effect if it is indecent: good

[8] Let us take an illustration from *Punch*: *Wife (in bed, to husband, a special constable, who is hurriedly dressing himself at* 3.45 *a.m.)*:

humour may be restored to quarrelsome conversation by a *risqué* anecdote. And in jocular intimacy the strange may be familiarized by the reversal of the meaning of a word, for contraries have an element of sameness. 'We've had a *lovely* day: it's poured all the time.' Naïve little pleasantries of this sort contribute much to the good humour of life. Indeed the revulsive susceptibility of our psychic force is one of the distinctive features of humanity. Its elasticity infuses life with a cheeriness which is man's peculiar possession. We are, each of us, attended by a jester to enliven the dull round of convention—a Spirit of Fun,[9] which can even ridicule incongruities in ourselves, and turn our own cherished dignity into laughter.

Finally of *intellectual* pleasures and displeasures. These are the consequences of interaction between the brain and psychic energy. Curiosity is attended with a feeling of exhilaration because it is expansive. Its subconscious resistance to mental confusion arises from the incompatibility of its expansiveness with this contractive condition. But a far greater pleasure attends the triumphant success of resistant reasoning that is *consciously* stimulated by doubt or hesitation. For the consciousness of the conflict intensifies the pride of victory. So one of cultivated mathematical tastes can derive intense pleasure from the solving of problems that offer nothing in the way of practical advantage.

We have distinguished pleasures and displeasures *in phase* according as their stimuli are presented in actuality, in recollection, or in imagination. Any stimulus may be recalled in memory and will repeat its effect,

'Where are you going?' *Husband*: 'Air-raid duty, dear.' *Wife*: 'Well—don't let the cat out.' In experience there is no connexion between the stress of an air-raid and the vagrant habits of the cat. But there is a connecting relationship: the opening of the house-door is the cat's opportunity. Accordingly, our surprise is hardly excited before it is allayed. And the revulsion is enhanced by bathos. For the cat is on a far humbler plane than an air-raid.

[9] In his pathetic lines to his departing soul, the Emperor Hadrian laments—'Nec, ut soles, dabis iocos'.

in more or less degree as recollection is vivid or faint. We are thrilled by the remembrance of an exploit, distressed by one of a humiliation. By nervous association memory may even produce a physical effect: it may actually reproduce the conditions of the sense-organs which accompanied the impressions that it recalls: the recollection of food will stimulate an appetite.

Imagined stimuli may be equally powerful. They may excite us physically, spiritually, emotionally, or intellectually. Their physical effect is generally less vivid than that of sensation in that it is unaccompanied by the excitement of the sense-organs which is involved in actual impression. Antarctic explorers have found that hardship may be alleviated by the planning of imaginary banquets; but their fancies can seldom have been so realistic as to give them actual sensations of taste. Yet even this is not impossible. For if fancy is very active, its images, through a reversal of nervous associations in sequence, will produce the conditions of the sense-organs that have accompanied similar impressions in the past. Hallucinations are sensory conditions that are stimulated in this fashion—from within instead of from without. And imagination may augment the effect of actual sensation: it is its *decorative* effect that enhances the savour of food when appetite is keen and the pleasure of satisfaction involves strong psychic excitement. In their spiritual and emotional effects imagined may be as potent as actual stimuli. Fancied slights may depress us as heavily as actual rebuffs: we may fervently adore our ideal of beauty and afford ourselves actual emotional delight by 'building castles in Spain'. Emotion that is fancifully excited becomes extraordinarily complex when it is *pitiful*—when it takes the form that is called 'pathetic' or 'touching'. It is pleasurable because pity involves psychic expansion. But, being stimulated by images of failure, distress, or disillusionment, it may bring tears to the eyes. Thus one may weep over visions that recall the dreams and hopes of his childhood, and still be emotionally pleased by them. The term 'sentimental' is commonly used to express this strange mixture of pleasure and sadness. Finally, imagination may afford intellectual pleasure: we may enjoy the triumph of understanding a difficulty by fancying that we comprehend what, in fact, is incomprehensible.

Emblems, such as toys and pictures, present imagined stimuli 'ready-made'. And since the effect of an imagined personality is greatly enhanced if we are impressed by conduct and facial expressions that are closely associated with feeling in ourselves, the theatre affects us more strongly than fiction. Its attractions are manifold. Our curiosity is aroused by its incidents: our admiration by beauties of scenic dress and decoration and of the feelings that are portrayed. But the *dramatis personae* excite actual emotions of love, anger, respect, and pity; and, if our sympathy is kindled, we experience the feelings which they represent. Iago excites our anger, but we are impressed by his success: we dwell upon the misfortunes of Desdemona because they arouse our pity: Medea's vengeance may horrify us, but we sympathize with her in her abandonment, and admire her forcefulness. And when the characters or incidents are distressing, we can find pleasure in the thought that they are, after all, unreal. When, however, feelings run high, the unreality of the drama may be obscured: an audience has been known to storm the stage and execute its desire for justice upon the villain of the piece. So strong is this desire that few novelists or playwrights will venture to bring their heroes or heroines to an unhappy ending. Indeed the justice which is dispensed in fiction is not the least of its attractions.

CHAPTER XVII

EXPECTATIVE AND EXPLORATIVE BEHAVIOUR

We are now concerned with voluntary conduct as opposed to that which is forced upon us by instinct, by emotion, by automatic or subconscious resistance, by imitation, and by habit. Its distinctive feature is that it profits by experience. It is the practical consequence of the mental process of reasoning. The will, as we have seen, comes into play selectively when we are drawn by different pleasures or repelled by different displeasures. We balance the competitors, and choose to pursue or avoid the greater through recollections with which their intensities are associated. This deliberative effort would be impossible were the mind not steadied by psychic resistance. But it is the recollection of pleasurable or displeasurable consequence that determines our choice. This is obviously so when the objects of our deliberation are actually present before us, as, for instance, in choosing between two dishes at table. It is less clear when they are possibilities of the future. Our choice is then moved by expectations. These, we have seen, are appetitive anticipations that stimulate search for or avoidance of something that is unperceived. In this peculiar feature they resemble the appetites to which, it appears, they owe their evolutionary origin. Expectations affect us as pleasurable or displeasurable possibilities. But we do not always perceive that our conduct is determined by the attractive or repellant force of pleasure or displeasure because these consequences may be sought through causes which in themselves are colourless and may even be unpleasant. One may, for instance, with an effort, deny himself afternoon tea to give the appetite a sharpness that will increase the pleasure of his dinner.

When we are attracted by an expectation we pursue it, when repelled, we avoid it. Expectative behaviour is, then, voluntary pursuit or avoidance. These motives are evolved from instinctive propensities to approach or grasp the favourable, and to recoil from the unfavourable, which are essential for the satisfaction of the appetites and for self-protection. They become conscious as desire and aversion. We cannot pursue the unpleasant or avoid the pleasant, unless their immediate effects are contradicted and outbalanced by remoter consequences. If we refuse to take rest when greatly fatigued, we have in view a greater displeasure. If we deliberately encounter danger or difficulty, a pleasurable or displeasurable consequence entices or supports us.

We need not enlarge upon the expectative effect of *physical* pleasures and displeasures, as of food and hunger, lust and pain, comfort and discomfort. Their pursuit or avoidance is the primary motive of voluntary behaviour, and to some men the attraction of alcohol is hardly less compelling. The appetites are distressful; yet, as we have seen, we may seek pleasure by sharpening them. The voluntary avoidance of pain and discomfort is the motive upon which the law relies for the repression of crime. It is the *raison d'être* of the medical profession. We search indefatigably for the causes of our ailments, and, having learnt them, anxiously keep our distance from them.

The aesthetic pleasures that arouse our admiration attract us exactly as do physical pleasures. If the excellences that delight us are attributed to sensory impressions, we pursue these impressions through their stimuli, as in ordering an excellent dinner, attending a concert, or visiting fine scenery. The excellence of dexterity in action or speech may invite us to attend a cricket match or a political meeting. Aesthetic pleasures of sight may also be sought through emblems:

we may please ourselves by looking at pictures of beautiful women. Excellences of feeling, other than of our own feelings, can be appreciated only through emblems, such as are afforded by the theatre, and by such pictures, statues, and compositions as move us, not through dexterity of execution or decorative beauty, but by the representation of feeling. The emblematism of Art is a means by which, through perception, we can obtain aesthetic pleasure from ideas of emotions.

Romantic, egotistic, and idealistic pleasures are imaginative, and cannot be 'pursued' in the literal sense of the word. But we appear to pursue them when we seek to express them in our conduct, since our expression of them *follows* their effect upon us. Expression is imaginative behaviour, and romantic, egotistic, or idealistic conduct is, therefore, strictly speaking, not expectative. But the effect of ideals upon us being *practical*, resembles that of expectations. We manifest the influence of romantic pleasures by behaving romantically; of egotistic pleasures by self-consciously endeavouring to win a feeling of excellence by overcoming our physical selves, or by decorating ourselves; of ideals by striving to act in accordance with them.

So we pursue such ideals as those of Dignity, Liberty, and Justice, referring to them as our 'principles' of action. They are attractive because they are derived from the power that is the cause of pride. Farther back in the sequence is the resistance that is the cause of successful power. This is idealized as Magnanimity, and in pursuit of it we may welcome[1] difficulties or temptations, and even seek them in order to resist them. An ideal of Unity pleases because it extends one's own individuality: it is the fountain-head of Sympathy. Ideals are of incalculable service in enabling

[1] As expressed dramatically in Browning's lines:

Then welcome each rebuff
That turns earth's smoothness rough,
Each sting that bids nor sit, nor stand, but go!

us to pursue psychic pleasures otherwise than through their material stimuli. Being psychically inspiring they sway the behaviour of individuals and nations with a masterful influence. They may breathe into man a disinterested courage—a magnanimity—that seems to be indifferent to material consequences. But they may be actively injurious. They may inspire him with an unreasoning prejudice or revengefulness which renders him indifferent to the sufferings of others. They have influenced men's history for good and for evil. They have established systems of religion and morality which, beginning in the magic superstitions of caste, have flowered in the spirit of Christian sympathy. They have led man into revolt against his natural instincts, and have fired him with self-repressive ambitions that have ranged from monastic asceticism, through puritanism, to the noblest dictates of duty and discipline. As antipathies they have hurled one nation against another in wars that have almost destroyed civilization; as sympathies they have raised visions of universal peace and brotherhood. They are the most potent instruments of persuasive politicians. They may be endorsed by experience, and in this case they are persistent. But otherwise they fade away. Expediency comes into play—the intelligent balancing of economic values which smiles at self-sacrifice as a good in itself, and indicates the line of least resistance as the most prudent course. Riches, it is claimed, will give dignity as effectively as idealism, and will give much more besides.

Accordingly the pursuit of *economic* pleasures, or values, is hardly compatible with idealism. It is less admirable, and does not raise man to so high a level; but it will not drag him down to such depths of passion. Like the appetites, from which it has evolved, it grows in strength with indulgence, so that it has become the characteristic feature of present-day civilization. The economic value of money is the stimulus that has led to the immense development of manufacture and trade. It is profoundly influencing the reproduction of civilized mankind by discouraging early marriages and large families. Under its lure man might become as soulless and mechanical as are the insects,[2] were it not for occasional outflashes of idealism.

Money-making in expectation is 'business',[3] as opposed to pleasure, for, its pleasures being remote 'advantages', it is uninspiring. Work for money's sake is unexhilarating. And generally it involves the irksome persistency which is called 'taking pains'. Continuous labour involves fatigue and is naturally distasteful to mankind, as may be inferred by the dislike with which children and savages regard it. When it becomes habitual, it is less tiresome, because it is less fatiguing; unconscious movements take the place of voluntary resistant effort. But habit is uninspired by pleasure, and it is a trite conclusion that industry sacrifices present enjoyment in the interests of a future which may never come. Civilized life is typified by the taking of medicine for the improvement of health. The lives of most men are spent very largely in doing what they dislike in order to obtain consequences which they like. The position may be worse. One may do what he dislikes in order to avoid what he dislikes still more. This is slavery. But it is the condition of vast numbers in countries which make a boast of freedom.

So it comes that rest or a holiday is delightful, until expectations of future consequences reassert their influence over us, and we feel obliged to recommence our instinctive pursuit of them. Fiction gives the relief of play. There is no one who is not charmed by an escape from industrious routine into an atmosphere of romance where life is thrilled by such emotions as love, courage,

[2] We may make the antithesis that the unconscious evolution of co-operative industry culminates in the bee, its conscious evolution in the *bourgeois*.

[3] 'What *business* had he to be there?' means—'What were the future expectations that led him?'

or pity—is, in fact, storm-driven by the heroic instead of being led in tow by the commonplace.

It is the development of economic pursuits that differentiates modern from ancient society. There is a gulf between present-day motives and those depicted in the Bible or the Homeric poems, and, speaking generally, there is a similar contrast between the objects of life in the West and in the East. The attractiveness of self-renunciation for its own sake—typified by asceticism and magnanimity—has waned under the magnetic influence of economic values—a change which the classical poets deplored in their regretful references to the 'Golden Age'.[4] They were nearer to it than we are, for in their days money was a comparatively recent invention. The poor are less tempted to set economy before magnanimity, since their expectations are uncertain and do not fascinate them. The generosity of the working man is astonishing and admirable—however ridiculous it appears from the economic standpoint. In India the ascetic still ranks far above the millionaire. If we class magnanimity with the highest of virtues, we can understand why the Gospel doubts whether the rich can attain the kingdom of heaven. And, although its attraction may have weakened, we continue to rate it highly: it commands universal respect, whereas another's prudence leaves us cold, and we cannot help despising the profiteering spirit. It *pays* a government to be magnanimous, if it would win the admiration and respect of its people and of other nations.

Magnanimity does not, of course, stand alone in the shadow into which it is pressed by economic interests. Idealistic and imaginative pursuits, in general, suffer the same eclipse. The chivalry and romanticism of the

[4] Thoughts of profit are focussed by money, but occurred, of course, before money was invented. In one of the noblest episodes of the *Iliad*—the encounter between Glaucus and Diomede, when the war-spirit was subdued by recollections of family friendship—the poet cannot restrain a smile at the *imprudence* of Glaucus, who, in fraternally exchanging arms with his opponent, gave gold in return for brass, in magnanimous indifference to economic values.

Middle Ages invest those times with the glamour of melodrama. Money was then scarce, and mankind were not tempted by expectations of expectations to sacrifice to the future the imaginative charms of the present. In those days idealism mastered mankind : it now serves politicians as a useful means of kindling enthusiasm for utilitarian enterprises.

We can pursue *emotional* pleasures by deliberately indulging passions that have been excited, as, for instance, by schemes for meeting one whom we love. But we cannot *will* the onset of an emotion which includes physical elements, and is not wholly psychic, although we can invite it by imagining stimuli which, having traits in common with actual stimuli of the past, can reproduce their consequences associatively. Thus one can fancy himself into love. Ideals are imaginative stimuli, and accordingly feelings of loving-kindness may be induced by thoughts of human fellowship. We can, however, avoid any emotion by shunning its stimulus, as, for example, by renouncing the society of one who attracts us or irritates us. The resistant emotions can be *willed* in themselves, since they are not led by external stimuli, and can be isolated from them. Dangers, difficulties, and even bloodshed[5] may be sought because they entail a resistance the consequence of which is pride.

Pride is a spiritual pleasure, evolved from a psychic revulsion, and leads us to the consideration of expected pleasures of this kind. Dignity and honour are sought through symbols, or through the respect of others. We strive to obtain symbolic distinctions and to win respect or popularity by our dress, belongings, manners, and conduct—a motive which has contributed very materially to the development of cleanliness, decency, and

[5] The philosophical Sanskrit treatise called the *Bhagavadgita*—the favourite reading of the cultured Hindu—opens with a conversation between the hero and his charioteer, as they drive between the lines of two opposing armies, his own and that of his cousins. He deplores the strife which is about to set blood relations in deadly conflict. The charioteer reproves him, pointing out that life is a passing dream, whereas glory is an enduring reality.

modesty. And we spend much—it may be, most—of our time in searching for the revulsive pleasures of amusement. Joy and sorrow are also primitively revulsions, but they become states of psychic expansion or contraction that are associated with the effect of certain stimuli, and are pursued through these stimuli.

Expected pleasures of the intellectual kind are pursued by *explorative* behaviour. The thought by which this is actuated is primitively stimulated by subconscious curiosity. But this is not a motive which would carry us very far, and the exploration which has unveiled the truths of science has been urged by expectations of the pride, or economic value, of success, or by an ideal of Truth which demands expression by discovery. Explorative activities are illustrated by reading, by travel, by scientific investigation, and by the observations and experiments that are made, not only in laboratories, but in the ordinary course of intelligent life. It is through experimental trials, as will be shown, that has evolved almost everything that distinguishes our material life from that of the brutes—our methods of cookery, our clothes, houses, and furniture, and even our manners. When experiment concerns itself with the nature and constitution of our environment, it is an instrument of Science.

The contrast between ancient and modern civilization is heightened by their attitude towards scientific investigation. The philosophers of Palestine, Greece, and India occupied themselves with the mind and spirit of man, which seemed to them far more interesting and important than the constitution and laws of the outside world. They formulated their discoveries in writings the practical wisdom of which is not prejudiced by the belief that psychic and mental activities were of spontaneous origin, or were inspired by the Divine. In those days no other conception was possible, since the processes of Nature remained quite obscure, and had not disclosed the law of evolution. Subject to this limitation their study of mankind was exhaustive. The Bible is a treasure-store of truths—a far more reliable guide to human conduct than the one-sided theories of the present day which take only economic considerations into account.

External nature, on the other hand, was regarded from an imaginative point of view. Its phenomena seemed irreducible to law, and were generally attributed to Divine intervention. They gained in picturesqueness what they lost in truth: it gives more pleasure to figure the sun as Hyperion than as a globe of incandescent matter. But so long as we attribute the wonders of Nature to supernatural agency, our curiosity is not aroused, and our environment does not present itself as a subject for scientific study. Moreover, some of the acutest reasoners of those days were prejudiced against the study of Nature by a vivid realization of the symbolic character of our ideas—by a conviction that they were an untrue presentment of actualities. Sensation appeared to be an illusion, except in so far as it concerned oneself. Pressed to extremes this view denied all philosophic interest to the behaviour and fortunes of mankind in the aggregate; and it is a remarkable fact that the wealth of classical Sanskrit literature does not include a single work on history.

Accordingly such theories as concerned themselves with physics were limited to the ultimate constitution of matter, and, being unchecked by experiment, were merely speculative. But scientific observation bore its firstfruits in Aristotle's studies of natural history, and in the histories of Herodotus and Thucydides, which will remain for all time models of dispassionate reasoning inquiry—of the exploration of truth for its own sake, all the more remarkable because they were written in the infancy of the scientific spirit.

One reason for the disregard of science in those days was that its discoveries were of no economic advantage. It is only with the development of manufacture and trade that scientific knowledge has acquired a money value. Moreover instruments of precision were not available until industry was organized upon a mechanical basis; and in providing them, manufacture has contributed very materially to the progress of science. Economic or applied science appeals to our admiration less forcibly than pure science, because it is the consequence of a physical stimulus, not of the resisting power of curiosity. But the hope of gain has undoubtedly encouraged the exploration of Nature, and we may infer that the development of science owes something to the invention of money. This does not imply that science is mercenary. Truth, for its own sake, has always been an object. But it gains in attractiveness when it is economically profitable.

It may, perhaps, be questioned whether the discoveries of science have added to happiness. For they limit imagination, and this is an abounding source of delight. As attributes of the Divine, the beauties of Nature are more wonderful than as illustrations of evolution. Childhood is the most imaginative, and the brightest, of ages. Moreover, imagination affords comfort and confidence by suggesting stimuli for faith. But imaginings may also be terrible. Savage life is overshadowed by dreads of the Unseen: apprehensions of hell-fire may render one miserable. And if the joys of childhood are keen, its griefs are also poignant. Science is disillusioning. But so is daylight. Yet we prefer our waking experiences to those of our dreams.

CHAPTER XVIII

APPRECIATIVE AND IMAGINATIVE BEHAVIOUR

THE phases of behaviour which we are about to discuss are not dictated by a practical object—the achievement of an expectation, or the discovery of the unknown. They simply express or manifest the excitement of pleasure or displeasure. The primitive function of expression is to relieve emotion by turning it into muscular channels. These, it must be realized, lead to words as well as actions, since utterances involve muscular movements. Vague or thoughtless excitement vents itself automatically in smiles and frowns, cries and gestures. But, when thought intervenes, the liberation becomes that of the ideas that are stimulating. Accordingly the *appreciation* that is involved in a like or dislike is shown by words or gestures which manifest it.

By a further very subtle elaboration appreciative pass into imaginative manifestations. We can follow the course of this development if we take careful note of the fact that nervous excitement relieves itself by converting feelings into *tangible* consequences—a feeling of pleasure, for instance, into a smile, that can be felt, in the sense of *touching*. The muscular *consequences* of excitement become *causes* that give a tangible or concrete character to excited thought. Movements evolve into motives. The lively images of enthusiastic reflection reinforce the excitement that produced them, and express themselves in behaviour which, so far as is possible, imitates them. It appears, then, that our imaginative faculties arise from an acuteness of psychic sensibility that subjects us to the excitement of which we are conscious as pleasure and displeasure. With increasing intensity there is increasing need of channels of escape.[1] Pleasure and displeasure have a *practical* evolutionary value as motives of expectative, or chosen, conduct. But their effect in stimulating imagination is merely *incidental*, however important be the part which imagination plays in our lives.

Appreciative behaviour expresses likes and dislikes, admirations and contempts, and also signifies their intensity. Its primitive forms are smiles and frowns, cries and gestures of applause or contempt—the 'tell-tales' of our feelings. These evolve into manners which express towards others the feelings with which they move us. Greetings are expressions of respect or affection. The social function of 'calling' is an approach of admiration or respect. Ceremonies express feelings by manners which are fantastically embellished by imagination. The gloomy pomp of a funeral signifies at once our sorrow for the dead and our sense of his importance. So with the feelings that are inspired by our own egotism. A sense of dignity is expressed by a stiff, or pompous, deportment; a sense of humiliation by a relaxation of the body muscles. Indeed, if we review such of our manners as are not instruments of expectative behaviour, we shall find that they are all appreciative expressions.

We most commonly, however, think of appreciative expression as verbal—as praise and blame. These are critical or emotional according as they involve reference to a generalized standard of excellence or morality, or are dictated by the emotion of the moment. They may take the forms of compliments and snubs. We may praise or blame ourselves as well as others, since we appreciate our motives and actions by the self-approval

[1] Following this conclusion, we may, perhaps, suppose that the brilliant colours of birds and insects originated as automatic and hereditary expressions of nervous excitement, that were not eliminated by the struggle for life. Indeed the pressure of nervous excitement may have been the origin of variations in form, as well as in colour, experience controlling but not initiating their evolution. In this case nervous excitement would have an evolutionary, or creative value, quite apart from its effect in developing expectative behaviour.

or disapproval of conscience. Self-reproach involves regret, and when this is acute it becomes repentance, which may express itself in a complete change of emotional or expectative pursuits.

To praise is to 'magnify'. This introduces us to a curious instance of the collaboration of appreciative and imaginative activity. We have seen that one of the effects of imagination is to *decorate*: we heighten ideas of excellence or inferiority by exaggerating their traits. Magnification and vilification are exaggerative embellishments. Thus one who is boastful appreciates his own virtues in magnifying them, and jealousy commonly expresses its dislike by vilifying its object. When we admire or respect an ideal we are pressed to magnify it, and to give practical effect to its magnification—a motive which invites the votaries of a civil or religious ideal to spread its influence enthusiastically by conversion, or even by conquest.

Politics, in so far as they are not influenced by a practical purpose—that is to say, by the 'opportunist' pursuit of public or private advantages—are the magnification of certain ideals, that are adopted as 'party cries'. The spirit of emulation is introduced, since different parties adopt different ideals. But it is the idealism of politics that renders their course so interesting. This is imaginative, as is sometimes recognized in styling them 'a game'. Accordingly, politicians must have principles to attract attention. It is only so long as we imagine ourselves to be building a 'New Jerusalem' that we are enthusiastic over the provision of houses for the poorer classes. It follows that although an opportunist or coalition government may be practically the most useful—and may indeed be necessary for salvation in times of danger—men soon become tired of it when the crisis is passed.

We turn to Imaginative behaviour in its typical phases. These have already been touched upon in chapter ix when discussing imaginative thought, since imaginings can hardly be divorced from their expressions. Imaginative behaviour gives tangible actuality to pleasing or displeasing ideas that have been tangibly figured in imaginative thought—that is to say, it expresses them by fashioning them, or by acting, speaking, or writing them, since these processes all involve touches.

Now, a pleasing or displeasing idea, if it affects us physically instead of imaginatively, will arouse an expectation and will incite us, not to expression, but to practical pursuit or avoidance, and it appears that its influence, this way or that, depends upon our psychic susceptibility. If the psychic excitement aroused by the idea is acute, it stimulates expression at the expense of expectation—that is to say, psychic influences overpower those of the physical plane. Accordingly, the stronger is the psychic sensibility the more imaginative will be the behaviour. Experience bears this out. Children are exceedingly susceptible to psychic stimuli, and in their lives imaginative 'play' ousts 'business' altogether. This susceptibility persists throughout life in those who possess the artistic or 'Bohemian' temperament. They live in the present, and are not obsessed by the expectations that arise from the influence of ideas upon our physical nature. But, speaking generally, business gains upon pleasure as experience grows and the scope for reasoning widens. Indeed this conclusion epitomizes the drift of civilization. A people matures —or 'declines'—when the spirit of imaginative creativeness is lost amidst a multitude of sensual and economic interests—when idealism and romance are atrophied in the shadow of instinct. What imaginative activity remains is merely decorative, or reverts to critical appreciativeness. And faith shrivels into a belief in the protective influence of caste and of magical observances.

One in whom the imaginative, or artistic, temperament persists is 'forced to express himself'. He may, it is true, be influenced by more practical motives—the gratification of vanity, winning of renown, or the earning of money—an incentive under which he may drift into the composition of 'pot-boilers'. But these are subsidiary influences. Pure art is content to express for expression's sake. Shakespeare was evidently quite indifferent to notoriety, and did not even trouble himself to ensure that his dramas were correctly recorded on paper.

Purely imaginative behaviour may be either playful or artistic. In play, imagined ideas are expressed as they suggest themselves: in art, their expression is directed by certain rules, or canons, of excellence. Children's play is obviously a form of 'acting'. It is the germ of the dramatic art—the simplest form of imaginative expression—and we refer to the theatre as 'the play' *par excellence*. We do not discard play when we outgrow childhood. Our social life is dressed in a fanciful tissue of decorative pretences. Our games express imaginative ideas of emulation and victory. To flirt is to play with a living emblem of love.

We realize that art is a serious form of play, and speak of the artist as 'playing with his subject'. It is serious because it must conform to certain standards. Dramatic art, for instance, must present a consistent plot and must be 'staged'. The standards are more or less conventional, and hence there is little similarity between the symbolism of European and Chinese acting. The plot which is presented is itself a manifestation of imaginative activity, expressed by its author in writing. His work is a 'composition', since it expresses imaginative ideas through incidents of experience. He again must follow certain rules: there were times when he could not present more than three characters, or violate the 'three unities'. Authors, whether poets or writers of essays, history, or fiction, must conform to established methods if their work is to find favour with the taste of the day. The musical composer is similarly fettered. An artist may break from convention by inventing a new style. But he has generally to wait for popular success.

Imaginative expression by fashioning an object, as in painting, sculpture, and architecture, is of much later development than the drama and poetry. During all time it will be difficult to surpass the excellence of the Homeric poems; but when they were composed, the plastic and pictorial arts of Greece were in their rudest infancy. To fashion, as we have seen in chapter x, is to imitate an idea in material. The idea may be decorative or creative. To decorate is to magnify by embellishment, and is therefore an imaginative form of appreciation. The ornamentation may be simply imitated from nature, or be suggested by imaginative ideas of excellence. The acanthus-leaf border illustrates the former: the key-pattern the latter. Its attractiveness is increased by rhythmic repetition, since this makes a physical appeal to us. Imaginative handiwork is, as we have seen, of much later origin than the drama and poetry, and the specimens of antique art that have come down to us are generally uncouth or grotesque. But we must except the sketches of animal life which the palaeolithic cave-dwellers of eastern France scratched on pieces of bone or horn, and on pebbles. These are of amazing vivacity, although so ancient that some of the animals which they figured have since become extinct. In those days the reindeer, the mammoth, and glutton flourished in southern Europe. A decorative style, once established, is maintained by fashion—often through such a process of degradation as deprived the acanthus leaf of its natural elegance.

Handiwork that expressed creative imaginings made its earliest efforts in the fetish and the idol—the former a purely fantastic emblem of a mysterious force, the latter imitative in its resemblance to the human form. The symbols of creative art must be anthropomorphic to affect us emotionally, for we can only realize another's feelings through his expressions of emotion. Force can be obtained by exaggerating expression: this was the aim of Byzantine art, in its stiff representations of the dignity of holiness. But the more nearly expressions resemble those of experience, the more vivid is their effect in recalling ideas of feeling, and modern art is convinced of the power of Naturalism. But from time to time protests arise: such are Futurism and Cubism. In creative, as in decorative art, style tends to become conventional, and expression may become rigidly formalized as it was in ancient Egypt. Moreover, creative art is exposed to another danger. As execution improves with experience, the pride of technique may engross the artist's attention, and he may over-decorate his phantasies—so bedecking the creatures of his brain that their vitality is smothered by their clothes.

Art in its earliest stages appears to have been very largely influenced by 'magical' reasoning. But, with a sharper appreciation of differences, magic lost power, and the artist passed from the magician into the 'maker', who fashions new worlds into which others may enter and wander at pleasure. Not only does he give an imaginative vitality to the passing features of our environment and ourselves, as when he embodies natural or emotional beauty in a picture or a statue, or the canopied shade of a tall avenue in a Gothic cathedral: he endows his creations with decorative beauties of their own. So the dramatist and novelist transfigure the various traits of human nature into living characters, round which is woven a decorative tissue of incidents and conversations. The poet adds the charm of rhythm to other imaginative embellishments.

And what of music? Its effect upon us is threefold. The rhythm of its progressive intervals—the beats of its *time*—appeals to a susceptibility which seems to lie deep in our physical nature. The atmospheric vibrations which are translated by the nerves into *notes* (and combinations and successions of them in accord or discord, and in major and minor keys) touch us psychically, independently of the brain, producing, it appears, nervous conditions which have some actual resemblance to the emotional. And, lastly, we are aesthetically impressed by the beauty, or excellence, of its harmonies and melodies. Its physical and psychic effects are felt by us naturally, subject of course to differences in nervous susceptibility. But ideas of musical beauty, representing conventional standards of excellence, may vary from time to time, and from nation to nation. Accordingly the canons which guide the composer are in great measure arbitrary. Within, however, the limits imposed by them, his imagination enables him to conceive of chords and progressions which he has never heard—to manifest his creative and decorative talents by forming new combinations of experienced impressions. In fact he uses time and sound as artistic materials: they are *tangible* in that they touch our senses. But his skilful utilization of them is primarily stimulated by his delicate appreciation of the similarity between the effects of music and emotion, which renders melodies and harmonies a natural expression of emotional excitement—as its song is to a bird. And to those of acute musical susceptibility his compositions give sentimental as well as aesthetic pleasure. A musical theme may be compared to the words of a drama, which call up by association a succession of feelings. Words, like notes, are *sounds*. Music can recall emotional feelings associatively. But, association apart, it produces feelings that are akin to emotion.

Accordingly, imaginative behaviour emblematizes ideas that have been formed under the influence of emotion, and decorates the emblems which it creates. Emblems are idea-signs having something in common with the ideas which they represent. Upon others they act as stimuli, exciting appreciation, and suggesting imaginative thought through a trait that has been associated with experience. By emblematizing a horse one child leads another to imitate its driver. A statue of Venus shows us a fantastic personification of the power of love, and arouses imaginatively our like and our admiration. Church-bells do not merely notify the hour of service: their music is a message of good tidings. Some idea-signs, as we have seen, are *symbolic*: they are linked to the ideas which they represent simply by memorial or conventional association. The idea-signs used in speaking or writing are generally of this kind: most words, in themselves, are meaningless. The expressions of play and art, on the other hand, are imitative emblems: they are based upon some identity of trait between the idea and its sign. The resemblance may, however, be very slight. Children, it has been observed, will adopt anything which is longer than it is broad to represent a person. But they will not use a ball for this purpose except as an indication of position.

This is a very summary notice of a faculty which adds immensely to the happiness of life by distracting us from the constant harassments of the future. But it will suffice for the analytical purposes of this work. Before, however, concluding this chapter we may glance at the effect of imaginative expression upon behaviour which is primarily expectative.

We have seen that imagination may profoundly influence both practical thought and conduct. By fanciful samenesses reason is lured into the processes of magic; in the atmosphere of religion, curiosity languishes, and doubt allows itself to be stilled by dogma. So expectative conduct becomes self-conscious under the fascination of egotism, fanciful under the glow of romance, transcendental under the fire of idealism. There may come about a further development. Under the influence of religion, expectative conduct may be drawn towards morality by a peculiar stimulus. For, when moral precepts are yoked with religious doctrines, not only does their breach involve Divine retribution, their observance has the propitiatory effect of prayer, and need not be safeguarded by expectations of reward or of punishment. We regard morality as an essential feature of religion because Christ's teaching insisted so strongly upon acts of loving-kindness and sympathy, and because the religion of the Jews was pivoted upon their Law. But history and current experience show that there is no essential connexion between the two ; we need not go back to the Anabaptists to learn that religious fervour may actually dissolve the bonds of morality. If they rested upon reason, prayer and the higher morality would be discouraged by the results of experience. But, through imagination, their expression becomes an end in itself.

CHAPTER XIX

THE EVOLUTION OF METHODS

Our methods of conduct and speech have been transmitted to us by our elders; we receive them, so to speak, ready-made, and it is not easy to realize that they are the outcome of a gradual evolution, especially as the early stages of this process are hidden in the obscurity of the long past, and its fruits shed their evolutionary traces as they are handed on from generation to generation. Being mechanically assimilated by memory, our fashions of behaving and speaking appear to be instinctive and to be as inevitable as instinct. To the generation which witnessed their discovery they were notable inventive achievements, such as is the aeroplane of our days. But succeeding generations accept each of them as a 'matter of course'; we commit them to memory without scrutinizing them by the intelligence, and, if we discuss them, are concerned rather with their forms than with their origin. Memory has a tendency to stifle intelligence: we may observe amongst our acquaintances that one who commands a large store of quotations is disposed to use them in place of arguments. But if a man's body has been shaped by evolution, his conduct and speech must have evolved also. The fact that they evolve is illustrated by the changes which they undergo in each generation; fashions in manner, dress, amusement, and speech are, indeed, continually altering. Accordingly, in considering them we are concerned with their evolution as well as with their transmission by memory.

Let us endeavour to infer, from such facts as are within our knowledge, what were the circumstances under which man's conscious evolution commenced.

From the character of his teeth it may be concluded that his natural food is soft fruit and roots: he is unable to masticate hard uncooked grain or raw flesh, and his original habitat must, therefore, have been some locality where ripe fruit and roots are in season all the year round. He is unprotected by his skin against changes of temperature, and must have required a warm, equable climate. From both these points of view it follows that the conditions of his original dwelling-place must have been those of present-day Polynesia.[1] The use of fire, it may plausibly be conjectured, could only have been discovered by experiences of hot lava beds:[2] the Polynesian region is markedly volcanic. And we may reasonably suppose that his first essays in navigation were forced upon him by a depression of the land surface which broke up large islands into smaller ones, just as the occurrence of deep annual inundations has driven the inhabitants of Eastern Bengal into semi-aquatic habits, and has even developed a peculiar variety of rice which lengthens its stalks as the water rises—to seven or eight feet and even more. There has been such a submergence in the South Pacific.

The cause of his evolution was, as we have seen, the discovery of new pleasures or tastes. These might be either physical, as those of new foods, of cooked food, of shelter and comfort; aesthetic, such as those of decorative patterns and colours; economic, as the consciousness of future expectations gained intensity; or symbolic, as there were evolved the fashions of dress, manners, and art, by which pride is aroused and fancies

[1] It is a curious fact that, at an American exhibition of physical types of humanity, a Polynesian took the first prize. And it is to be remarked that we find in this region—in the Papuan and Sawaiori (Maori) races—representations of the two most divergent types of the human species. There is a good deal to be said for the Polynesian origin of the peculiarities of the so-called 'heliolithic' culture (such as circumcision, tattooing, and the *couvade*), which can be traced across America as well as Asia and Europe.

[2] A lava stream will conserve for many years enough heat to kindle a torch (Geikie's *Text-book of Geology*, p. 280).

are expressed. The *courses*, or lines, of conduct and speech, by which these pleasures are obtained, fall, as we have seen, into two classes. They are practical *means to an end* when they achieve an emotional desire, and when they are expectative or explorative. They are *ends in themselves* when they simply manifest appreciative or imaginative thought and feeling; in this case they may be distinguished as the 'sentimental' consequences of psychic, or spiritual, excitement. Such are our Manners, the activities of Art, and the observances of Religion. These are essential differences. But they are obscured because we so frequently act with mixed motives. Thus dignified manners may be used, sentimentally, to express dignity, or, practically, in order to obtain the respect of others: the work of an artist manifests his genius, but may at the same time be employed to earn money.

Our *methods* are means by which we achieve practical or manifestative behaviour. Ideas of them have been derived from instinctive and automatic movements, from assertive or tentative efforts of will, or from the observation of things around us. The spirit of conscious evolution has been man's curiosity, which incessantly urges him, not only to taste or manipulate, but to observe his surroundings. Its instruments have been imitation, reason, experiment, and accident. Imitation, as we have seen, is the expression of an idea by movements that have some trait of resemblance to it. We ordinarily think of imitation as the copying of another's actions. In this case we obtain from his movements the ideas which we express. But imitation may possess a more original quality. It may express an idea which we have conceived, irrespective of another's behaviour. We have distinguished the two phases as *submissive* and *expressive*. The former transmits methods: the latter may invent them. Human actions must have evolved before they could be transmitted. But man has always had before him the instinctive behaviour of beasts and birds, and, in savage or childish play, has delighted to mimic them. It seems probable that he learnt from them to climb and to swim, and he may be indebted to them for his primitive notions of moulding, weaving, house-building, and even of singing. This possibility will not seem far-fetched if we recollect that in our own time aviation has admittedly been copied from the birds. Through expressive imitation imagined ideas are expressed in tangible forms: in painting a picture an artist imitates fanciful ideas by materializing them on canvas.

Reason argues from a trait to a sequence of perceptive or reflective experience, or from a sequence to a trait. In experience a nest of crossed sticks does not fall down: a durable platform may, then, be constructed of sticks if they be crossed. It is a trait of the sticks that they are crossed: crossing may accordingly be imitated in fibre. The sequences of experience must of course be remembered in order to be utilized. The vast difference between the reasoning achievements of man and the lower animals is no doubt due in part to the fact that they have little memory for symbols. An animal so intelligent as the elephant can only remember a few words of command. It is, therefore, infinitely less susceptible than man to familiarities of rhythm. And, owing to a similar insensibility to familiarities of trait, ideas cannot be disintegrated into generalized traits: the attraction of sameness is not sufficiently strong to unify identical traits that occur in different ideas, and isolate them from their ideas, unless they are quite elementary in kind. Beasts and birds are, then, unable to elaborate definite ideas of a future, since expectations are appetitive anticipations of generalized consequences.

Experiment, as we have seen, is the expression of a tentative volition. Under the influence of explorative thought it becomes an instrument of investigation. It may be employed either to verify a hypothesis, or, tentatively, to explore possibilities. If a stone be held, the gesture of a blow be made, and the finger-grip be relaxed, the result is a *throw*, from which may well have evolved the use of a missile as an instrument for striking at a distance. A vast number of discoveries have been due to tentative experiment: the evolution of the motor-car has resulted from a long series of explorative trials.

It is clear from everyday experience that lucky accidents have contributed very greatly indeed to the discovery of methods. They may have occurred in casual experience, or in the course of acts of assertive or tentative volition. In manipulating a stone, it may be learnt accidentally that it crushed shells more effectively than the fist: by an extension of this trait, it could be employed to crush other things.[3] The experience which leads to accidental associations may be that of necessity or compulsion. Thus a dog associates 'sitting up' with food, and, if it be intelligent, may use it to obtain other favours. We may seem to rate too highly the value of accident as an instrument of evolution. But most men, on reviewing their lives, will find that accidents[4] have played a part of immense importance, often determining such crucial matters as the women they marry or their choice of a profession. It is then not unreasonable to suppose that it has also contributed largely to the evolution of conduct; and, as a matter of fact, some of man's most important discoveries during historical times are known to have been accidental. Was not gunpowder invented by mixing sulphur, nitre, and charcoal to 'see what would happen'?

Methods are of conduct or speech. In this chapter we are concerned with the former. These may be dis-

[3] Amongst the very earliest stone implements are roughly chipped masses of flint which must have been used in crushing. They were disinterred in the valley of the Somme by M. Boucher de Perthes, and given the name of 'coups de poing'.

[4] Hence purely personal incidents may cause political cataclysms—a fact which was more clearly appreciated by Herodotus than it has been by most of his successors.

tinguished as actions, instruments, or processes. But they are commonly used in combination, and hence the distinction between them will not serve as the basis of a precise classification. By *actions* are meant non-instinctive movements of the limbs, such as are used in swimming. *Instruments* are, in fact, extensions of the limbs: one who uses a stick to strike another adds to the efficiency of his arm: he 'arms' himself with it. *Processes* are the sequences of cause and consequence which we observe outside us and within us. We term the former 'natural laws': they include vegetable growth and the power of fire, steam, and electricity which is harnessed to machines. The latter include the shame which follows disgrace, and the dislike of punishment. They are utilized in discipline and government by being associated with certain acts, as their consequences. We extend the term 'law' to these artificial sequences. It would be beyond the scope of this book to attempt to trace in detail the evolution of civilized life; and it must suffice to show by some simple illustrations how the invention of actions, instruments, and processes has come about through the causes that have been outlined above.

First, then, of *actions*. These may be used indifferently for either practical or manifestative purposes. One may, for instance, climb a tree in order to reach a fruit or in play. But, if we regard their origin, they can generally be distinguished as primarily serving one of these purposes or the other. Of our *practical* activities some have an instinctive or automatic foundation. It has been plausibly conjectured that we express assent by nodding, and dissent by shaking, the head, because these are movements which accompany the swallowing or rejection of food, and there is a sameness of trait between swallowing and accepting, between rejecting and refusing. Our actions of aggression and defence are developments of striking, kicking, and biting—primitively automatic, or subconscious manifestations of antagonistic resistance. Their consequences being to injure an adversary, reasoning and experience would evolve from them other methods of causing hurt, culminating in the marvellously ingenious destructiveness of the late war. *Working*, on the other hand, is evidently non-instinctive: children and savages are naturally indolent unless their imagination is excited. Voluntary labour distinguishes man from the brutes by a gulf which it may seem impossible that evolution could have bridged. But it would naturally evolve from a growing appreciation of future consequences. Even civilized man works, in great measure, because he is pressed by necessity. To judge from savage life it is not instinctive with man to accumulate or *hoard* things. But the advantages of an accumulated surplus would become evident from accidental experiences: they are in fact its consequences which become expectations; and stimulated by these, the desire to possess would gradually evolve into a ruling passion—especially as possession gives the dignity of power.

Amongst our *manifestative* actions dancing is one of the most primitive of acquired accomplishments—indeed it is an automatic expression of ideas of rhythm. Singing is a more elaborate development, and, as we shall see, is impossible unless the vocal chords are practised during early childhood. It is not improbably a 'submissive' imitation of the birds. The 'acting' of children's play is 'expressively' imitative. The handclasp of greeting manifests the unity of kinship and friendship. The pride of success is accompanied by an involuntary bracing of the muscles—by a strut—and accordingly reason associates rigidity with a feeling of dignity; this evolves into a stately carriage, and into the crook of the little finger which is considered in some circles to dignify the lifting of a tea-cup. The reaction from dignity occasions a muscular relaxation which is at its maximum in the prostration of respect. The passage of this into forms of salutation strikingly illustrates the course of evolution. It is less irksome to lift earth to one's head than to level oneself with its surface, and the Oriental salaam obviously expressed this idea by a gesture which, modified in the interests of military precision, becomes the soldier's salute. Respect for another is also shown by such offerings of food (or sacrifice), praise, or service as we like ourselves, because the person whom we respect, being unified with us in interest, will share our likes. This appears to be the origin of the ceremonial that plays so large a part in religion and in manners.

Manifestative, or 'sentimental', conduct may at the same time be practical. Dignity, being associated with successful egotism, is at once manifested and pursued by actions which express resistance or exclusiveness—by magic *taboos* and fastidious observances such as those of the Indian caste system. Purity is a form of dignity: from it to personal cleanliness is but a reasoning step.

Passing now to *instruments*, the uses of a stick would be discovered in play; by an accidental improvement it would become a club. The idea of cutting would similarly arise from the manipulation of a sharp flint; it gradually extended its traits as man passed through the Stone, the Bronze, and the Iron Ages. The bow uses the elasticity of the branches that one bends back in passing through a forest. The original pick-axe was a forked branch, or a deer's horn; the primitive plough is a pick-axe drawn by animal instead of by human effort. It is not difficult to suppose that the drum and pipe were accidentally discovered in the course of playful manipulation. Dress, ornaments, and tattooing manifest the egotistical pleasure of peculiarity, and also serve as symbols of dignity. Respect is, therefore, shown by the removal [5] of clothing: Orientals take off the shoes, Europeans the hat.

The potter's art appears to have been suggested by

[5] Purchas relates of the Sultan of Melli that he would permit no women to enter his presence until they had discarded all their clothing.

the usefulness of gourds: they could be imitated in plaited fibre smeared with mud (a combination used by some birds in nest-building), and it is noticeable that the bases of primitive vessels are often rounded, not flattened, as would obviously be most convenient for standing them upright. So the flat surface of a leaf would suggest a fabric which could be made by beating out glutinous fibres into a sheet—an expedient still practised in Polynesia. Weaving, as already conjectured, may have been learnt from the birds. Building, it appears, has no instinctive foundation[6] in mankind. But the use of shelter from sun and rain would be easily appreciated: a hut reproduces the consequences of taking shelter under a tree, and there are many animals whose habits would afford ideas of construction. The instinctive industries of beasts and birds surround man with an exhibition of skilful fabrications, that have been perfected in the course of unconscious evolution, and literally represent the 'wisdom of the ages'.

We now turn to *processes*. These are sequences of which we think as causes and consequences. Perhaps the most important of all is the use of fire as a means of cooking and warming. By its discovery mankind learnt that cooked flesh could be masticated as easily as fruit, and he could migrate to localities where the climate and products were naturally unfavourable. Agriculture is the enlistment of vegetable growth, assisted by expedients the consequences of which could have been learnt by observation, and are still directing improvements in cultivation. The use of boats arose from the consequence that wood floats in water. That animals could be domesticated could be learnt from the keeping of pets; and this may be ascribed to imaginative influence, for one's fondness for an animal proceeds from indifference to its differences, so that it appears to be of the same kind as oneself.

[6] Unless the ourang-outan's nest of branches is made instinctively. One might, then, argue from it to man.

Means of locomotion probably evolved from the slave-litter. It is more dignified to be carried than to walk. When animals had been domesticated, the use of their drawing and carrying powers would be suggested by obvious identities. Riding is a comparatively recent accomplishment, and it is not so very long ago that a man on horseback was as remarkable as one now in an aeroplane. Finally, through an appreciation of identities in function, or consequences, assisted by trial [7] endeavour, man obtained the command of steam, electricity, and chemical forces which differentiates the modern type of civilization from all that have preceded it.

Amongst the processes which civilization has utilized are those of human nature. Exchange is a process of persuasion, based upon the fact that a man will part with a thing in order to gain another thing which he values more highly. If he has nothing else exchangeable, he offers his labour: thus he comes to work for hire. Before the invention of money, exchange was facilitated by the imaginative idea that the value of an object might be represented by a miniature emblem of it. The precious metals have, however, a value of their own which varies with their quantity; and the use of definite amounts of them, as coins, established a convenient standard of value and immensely facilitated the process of exchange. But money is something more. It has a high sentimental value as a symbol of power, and consequently it stimulates exertion quite apart from the comfort and luxury which it offers. And as a symbol of future pleasures, it has vastly increased the influence of future expectations. Its value rests upon the belief that others will appreciate it as highly as oneself. Confidence is secure if money is freely exchangeable for gold, since this is an object of universal desire. It has generally

[7] So Humphrey Potter, to escape the trouble of opening and shutting the valves of the primitive pumping-engine, connected them with the engine beam by suitable cords and catches, so that the beam, in its movements, should do his work for him.

been severely shaken by a debasement of coinage. But the value of money, like that of decorations, may be bestowed upon it by convention, and it is not inconceivable that mankind should be content to adopt an inconvertible paper currency as fulfilling its desires.

It may be observed that the fluctuations which occur in the value of money defeat mathematical analysis because they result in great measure from imponderable changes in popular mentality. Money, being the symbol of expectations, is dear if these expand, and cheap if they are contracted by an uncertainty of future prospects: or, put differently, the prices of commodities—or 'satisfactions' as they may be termed if services be included—are low in the one case and high in the other. Accordingly, that money should retain a stable value, the currency should expand and contract with the expansion and contraction of expectations. The use of credit gives it the required elasticity, for credit is the expectation of receiving money, and is therefore a substitute for it. Consequently, if the uncertainty of future prospects and their variations are within ordinary limits, prices remain normal, and credit maintains a balance between symbols and expectations at such rates of interest as are warranted by reasoned expectancy. But if future prospects become clouded or hazardous, this equilibrium is violently disturbed. For in this case the present is more attractive than the future, and money is diverted from purposes of investment to those of consumption—that is to say, from provision against future to the satisfaction of present expectations—and, being in excess for this purpose, it loses value, and prices rise. Normally this depreciation is arrested by the contraction of credit and rise in the rate of interest, which are also the consequences of uncertainty of prospects. If, however, the currency has been inflated by the State, this check is neutralized, and there may be an orgy of expenditure at very high prices. And, since the present engrosses attention, the rate of interest may run very high. The value of money may also be lowered by an exaggerated idea of future possibilities, leading to an abnormal expansion of credit —based, not upon reasoned expectations, but upon imaginative hopes, which stimulate speculation. In this case the currency is virtually inflated by extravagant borrowing. There is a reaction if these hopes are disillusioned, or give way to fears: credit contracts, the rate of interest rises, there is a commercial panic and a collapse of prices. An inflation of currency by the State may assist the community to withstand this disaster, unless, indeed, in the general demoralization, confidence in the currency also fails. It is, then, clear that the value of money depends very greatly upon emotional factors; and, as is well known, Sentiment holds court in the Stock Exchange.

The instrument of persuasion may be speech as well as money. So a government may support itself by the offer of expectations. But its control may rest upon a more powerful emotional sequence—the reaction of respect to the idea of another's power, and its consequence in obedience. This is the most primitive form of rule, and the most effective, since it is in great measure involuntary. Accordingly the revolt of democracy against reverenced authority exalts the influence of consciousness—of a desire for pleasure—as an evolutionary force. Another law of human nature is that one pursues the pleasant and shuns the unpleasant. Experience showed that security of life and property was unattainable unless a stop was put to robbery and private war. They could be checked by sufficiently unpleasant consequences. These took the form of punishments, imposed by a *law*. Civil government is an attempt to bring society under the control of rules that are reasoned additions to natural laws.

The course of these evolutionary developments must have been very slow, since reason has been consistently thwarted by the influences of faith, prejudice, and habit, and buds of growing culture have been incessantly trampled down and crushed out of life by war. Our manners and conduct are accumulated survivals from the ages which have passed since man first emerged from ape-like [8] simplicity. Yet we see their development marvellously abridged by our children: in a few years they pass through the stages which it has cost mankind aeons to traverse. They accomplish this by imitating their elders, that is to say, by materializing in movement or utterance ideas of movements and utterances which are gathered through the senses. By imitative obedience evolution is reduced to shorthand; the education of the individual is a lightning epitome of the evolution of the race. In his dependence upon training man differs radically from the lower animals. Their conduct, being largely instinctive, is independent of education and comes of itself. A puppy brought up apart from other dogs grows up with the manners and utterances of a dog; a cuckoo does not lose its distinctive habits or notes by being nurtured by hedge-sparrows. But if our theory of the evolution of conduct and speech is correct, it would follow that an infant which is brought up amongst brutish surroundings will not surpass the apes in its manners and utterances. That this is so can be demonstrated.

We know well that the language which a child speaks is derived from its associates. From the time of Herodotus anecdotes have been told of infants who, to satisfy curiosity, have been taken from their mothers, and committed to the care of she-goats, or women who were dumb. They all agree in the obvious consequence that in such a case children speak no language whatever. In regard to the effect of environment upon manners we can gather some curious and instructive information from observations that have been made of the so-called 'wolf-children' of India.

[8] Our opinion of man's conduct will depend upon our views as to his origin. As an angel, he is very disappointing; but he makes an astonishingly good ape.

According to ancient European and Asiatic tradition it has happened that children, carried off by she-wolves, have been nurtured, not eaten, by them, owing perhaps to an instinctive attraction such as that which a dog finds in human society. In India nine,[9] at least, of such 'wolf-children' have been known during the last eighty years, having been recaptured from their foster-mothers when they had grown too large to enter the wolf burrows. Five of them have come under reliable observation, and three of the cases are of comparatively recent date. All accounts agree that these unfortunates never learnt to maintain an erect position, or to speak; that their manners remained brutish and without a trace of decency, and that they even seemed to prefer the society of dogs to that of men. We pay for our brains by losses of instinct, and, judged by the physical plane, man compares very unfavourably with many vertebrate animals. And it seems clear from these cases that our brain avails us nothing if the plasticity of early childhood is wasted. It may be urged that the manners of all races of mankind are essentially similar, so far as the ordinary actions of life are concerned, and that they must, therefore, be innate. But all races of mankind are acquainted with the use of fire, and the same line of argument would prove that this knowledge is innate. The origin of human culture lies in the remote past, and there has been sufficient intercommunication to extend its primitive elements throughout the globe.

And we need not rely upon these curious abnormalities to be assured of the artificiality of civilization. History proves it. Under the effects of destructive wars, cultured man may revert to barbarism and lose all that has been gained by generations of ancestors. Classical civilization was extinguished by the tribes who broke up the Roman Empire, and during many centuries evolution was occupied in painfully endeavouring to redeem the past. With changed conditions, its steps have diverged

[9] Some particulars of them are given in Note D appended.

from its ancient footprints. But they have followed the same track. Between Lucian and Anatole France there is a gap of seventeen centuries. Yet the two are the same in ways of thought, although, during most of the period of ruin and regrowth which lies between them, there was hardly a man by whom either would have been appreciated.

NOTE D

THE first authentic account of wolf-children in India is given us by Major-General Sir William Sleeman in his *Rambles through Oudh*, published in 1849. The appointments which this officer held during his long Indian service afforded him peculiar opportunities of gathering information; he was drawn to the people by an active sympathy which opened their hearts to him; his interest was attracted by everything around him; and he was a careful and accurate inquirer. In the course of an official tour through the province of Oudh—at a time when it was still under Native rule—his attention was attracted by the large numbers of wolves [10] which infested some localities, and from time to time carried off young children. It had happened, he was informed, that the infants had not been killed, but been nurtured by she-wolves, and had been subsequently rescued when grown too large to enter the burrows. He heard of six cases [11] in which this had occurred within the memory of his informants. Two of these rescued waifs he saw himself, one of them having been for some years under the observation of an English officer (Captain Nicholetts of the 1st Oudh Infantry), who had given the boy shelter in his household. None of them showed any traces of human manners: the children ran on all-fours; they had no articulate speech; they preferred the society of dogs to that of men, and would readily permit dogs to share their food with them.

Eighteen years later, in 1867, another of these unfortunates was seen by my friend Mr. W. C. Benett (of the Indian Civil Service) in the Partabgarh Lunatic Asylum—also in Oudh Mr. Benett writes of his behaviour that 'his posture was semi-erect: he shuffled about on his "hind-legs", occasionally touching the ground with his hands: he made no use of articulate words: he had no sense of decency, and he preferred raw to cooked meat'.

More recently still, two wolf-children have come under observation in the Agra district. One of them was in the Secundra

[10] Within much more recent years wolves have been a veritable plague in India. It is within my own experience that some thirty-five years ago over a score of infants were killed by them during a twelvemonth in a single district of the Central Provinces.

[11] Some interesting references to these cases will be found in a paper of Sir Edward Tylor's published in vol. i of the *Anthropological Review* (1863).

Missionary Asylum from 1867 to 1895. His case has been described in a pamphlet which was written in the latter year by the Rev. C. S. Valentine, Principal of the Agra Medical Mission. He was discovered sitting in the company of a wolf at the entrance of a burrow, and was captured under the instructions of the District Magistrate and sent to the Asylum. During the years of his stay there he never attained a completely erect position; he never learnt to speak; his sole accomplishment seems to have been that he learnt to use his fingers in eating his food instead of 'wolfing' it. Another 'wolf-child' was in the Agra Government Lunatic Asylum from 1891 to 1913. The official record of his admission states that 'he was taken from a wolf's den'. I have been informed by the officer who was for four years in immediate charge of him that he could only walk erect on quite level ground; his position in walking resembled that of a monkey, except that his head was thrown well back; he never learnt to speak.

CHAPTER XX

SPEECH AND WRITING

SPEECH has an automatic origin in the cries and ejaculations which manifest and relieve psychic excitement. They issue spontaneously; but we are aware of them and can utter them voluntarily. They are the primitive germs from which words have evolved. For experience showed that they were of advantage: by them one might summon a friend or frighten an enemy. And, indeed, intercommunication affords a thousand pleasures and profits. We need not insist upon this point beyond remarking that, if we identify another with ourselves in kind or personality, his thoughts are as interesting to us as our own.

The evolution of speech has, then, been guided by an expectative appreciation of its advantages. But it has been impelled by the force of nervous excitement. Ideas that are exciting insist [1] upon being expressed by muscular movement. Speaking and writing are, in fact, gestures.[2] Most men have appreciated the relief which is given by an expletive. The insistent demand

[1] Helen Keller, the blind and deaf girl whose attainments were so marvellous, writes: 'The impulse to utter audible sounds had always been strong within me. I used to make noises, keeping one hand on my throat while the other hand felt the movements of my lips' (*Story of my Life*, p. 58). Darwin, in his *Expressions of the Emotions in Man and Animals*, contrasts the impatient wriggling and jumping of a dog when it sees its food-platter with its quietude when it has settled down to its meal.

[2] This is illustrated by two curious facts: firstly, that the area of the brain cortex (in Broca's convolution), which is concerned with utterance and writing, adjoins that which relates to movements of the arms and hands; secondly, that this area is normally confined to the left hemisphere of the brain, because this governs the right side of the body, and men are normally right-handed. With persons who are left-handed the area is situated in the right hemisphere.

of excitement for muscular expression is, accordingly, the primitive cause of the development of language, and, since this is also the cause of imaginative activity, language owes much to imagination. But its evolution has been directed by a sense of its practical advantages.

The development of language opens a vast field of inquiry in which multitudes have laboured under the spur of philological curiosity. We can attempt no more than to indicate generally the lines which it appears to have followed, and the processes of which it has made use. These may be distinguished according as they have been employed to invent words, or to extend, restrict, or modify their meanings. Of the first kind are (1) the simple or 'submissive' imitation of sounds, (2) the 'expressive' imitation of feelings, and (3) efforts of expression made in trial or in play. Words are *emblems* when they are imitative, *symbols* when they have originated in assertive efforts at expression. But emblems degenerate into symbols, as their original meaning becomes lost, and they are remembered instead of being understood. The meanings of words have been extended through similarities of trait or associations in experience, restricted by the separation of traits that were included in their primitive signification, and modified by the addition of syllables that were originally separate words.

First, of *inventive* processes. A word is a combination of sound and muscular movement which is attached to an idea and signifies it. If the idea is one of sound, it may be emblematized by simply imitating the sound, as in *thunder, squeal, crash.* If it is one of touch or sight, and a sound is associated with it in sensation, the sound may be used to represent it, as in *peewit.* For the association of the sound with an object in sensation suggests a name for it. Colonists in a strange land,[3] when they do not adopt the native name for an animal, almost invariably distinguish it by an imitation of its

[3] Farrar's *Chapters on Language*, p. 30.

voice. By a similar association a child expresses a cow by 'moo', and a horse by 'gee-gee', because this is the sound made in driving it. Sounds may be associated with bodily movements or with inanimate things, and can be used to signify them, as in *sob*, *cough*, *saw*, *hammer*, *tin*. Certain cries naturally express particular feelings, and appear to have contributed to the origin of the long vowels. *Ah*, *ih*, *uh* are the instinctive manifestations of surprise, pleasure, and displeasure: *eh* indicates confusion or distress, and *oh* attention. It is not altogether fanciful to suppose that the use of these sounds was guided by their emotional associations. There is certainly an emblematic meaning in *awful*, *cheery*, and *gloomy*: *ache* plainly imitates the involuntary cry of painful distress: *oh* is commonly used for the vocative, and in *odour* and *omen* seems to imply attention.

But its sound is only one of the elements which enter into a word. It comprises the muscular movements which form *consonants*. These are gestures[4] of the lips and tongue which are in themselves as independent of sound as are movements of the hands and fingers. The tongue has a tendency to mimic the actions of the limbs. Children when learning to write commonly imitate with the tongue the movements of the fingers. We can perceive that there may be a sameness of trait between our feelings of the movements of the tongue and lips and our feelings of movements of the limbs, so that movements of the one suggest themselves as emblems of the other. If we analyse the movements of the tongue and lips, in the utterance of such words as *blow*, *kick*, *jump*, *draw*, *shudder*, and *stop*,[5] we shall find that they possess a distinct analogy with the actions which they represent quite apart from any noise that may be made in uttering them: they are

[4] As illustrated in Note E at the end of the Chapter.

[5] So in *tendo*, *flecto*, the tongue expresses the feelings of drawing out, and bending back.

emblematic. In uttering *fly* and *flap* the tongue clearly mimics the movement of a wing; one can *feel* the significance of *push* and *pull*. Energetic actions are always preceded by a holding of the breath, and the words that express them commonly begin with a consonant that involves the arrest[6] of breathing by the tongue or lips; continuing movements, on the other hand, are signified by the use of sibilants or liquids, as in *see*, *say*, *run*, and *limp*. The tongue can also reproduce the feelings of external touches, and in *hard*, *soft*, *rough*, and *slippery*, we can detect similarities between external tactile impressions and impressions of the movements of the tongue and lips.

Words that are connected with their ideas by these identities of sound and feeling are forcible or picturesque, because they recall sensory impressions. In this respect English and German are better equipped than French, and are, accordingly, more vigorous, although, it may be, less precise. Words may possess identities of this kind although they do not owe their origin to them, and there is a tendency to turn accidental similarities of sound or feeling to account, since they increase vividness of expression. Homer uses this artifice with consummate skill.

It appears, then, that man's primitive vocabulary was more largely imitative than is generally supposed. But imitation has its limits. Sounds and touches may be reproduced, but one cannot mimic in utterance forms, colours, or thoughts. The assistance of another instrument was required. This was available in the efforts of trial which are stimulated by difficulty, or are

[6] Energy is also expressed by the lips and tongue in forming *w* and *sw* and *st*, as in *wake*, *swim*, *stretch*. *W* is formed by jerking the lips apart: when the tongue is jerked away from the palate, we have *ch*, *sh*, *j*, and *y*, letters that are also largely used with an energetic significance. It is probable that a study of 'glossomimetics' would account for the origin of many words that seem to be meaningless. For not only can the tongue and lips mimic the thumb and fingers, as Note E shows in detail. The thumb and fingers can imitate movements of the hands, arms, and legs.

urged by imaginative playfulness. Under the stress of an impulse to express, these can suggest an infinite variety of utterances, which will symbolize an idea although they may not emblematize it. In searching for proofs of this imaginative constructiveness we must nor disdain facts because they are trivial. It is a familiar experience that the children of a household ' play ' with words, and may invent a special ' nursery ' vocabulary of their own : we see this propensity illustrated in school and college slang, and very strikingly in thieves' *argot*. In his account of the Indians of North Brazil,[7] Bates gives a significant example of the playful coinage of new words : ' When Indians, men or women, are conversing amongst themselves, they seem to take pleasure in inventing new modes of pronunciation, or in distorting words. It is amusing to notice how the whole party will laugh when the wit of the circle perpetrates a slang term, and these new words are very often retained.' Words are signs, like playthings, and when the possibility of symbolizing impressions had once been realized, the imagination could fancy new words, and play with those in use precisely as children can make playthings out of their nursery furniture. It seems, indeed, that, until words were fixed by poetical tradition or by writing, language was in a constant state of flux. The tongues of modern Europe are crystallizations which emerged from these conditions of instability.

Our vocabulary is apparently more indebted to trial than to imitation. Most words in themselves are meaningless. Many of them, no doubt, possessed an emblematic significance which has been obliterated by changes of form. But, however this may be, the majority of words in use are merely symbolic.

Such, it seems, were the lines on which man's primitive vocabulary evolved. In some cases it still remains so meagre—as in China, for instance—that the same

[7] *A Naturalist in the Amazons*, i. 329.

utterances must serve to signify many different ideas, being distinguished by the tone of voice in which they are pronounced. It has been elaborated by subsequent processes in which meanings have been extended and restricted, and words have been amalgamated. The application of a word may be broadened through a sameness of trait—a metaphorical analogy. This has contributed vastly to the growth of language. Samenesses of appearance have led to such extensions as are simply illustrated by *larkspur*, less obviously by *gem*, which originally meant a bud. There may be resemblances between ideas that are derived through different senses: there is an obvious identity between the *tinkling* of a bell, and the *twinkling* of a star, and we frequently apply terms of sound to colour, as when we speak of hues as 'loud' or 'screaming'. We may imaginatively see identities between living and lifeless objects, and vice versa, as when we speak of the *arm* of a chair or of the *trunk* of the body. Analogies between perception and feeling are utilized very extensively in adjectives—'a sunny smile' for example. Mental processes may be identified with bodily action, as in *reflection* (bending back) and *attention* (stretching).

It is to analogies of this class that we owe the most abstruse of our words, and many of them show extremely intelligent inventiveness. *Deliberate*, for instance, most happily appreciates the similarity between balancing in the hand and in the mind; *consider*, the resemblance between earnest thought and 'star-gazing'; *difference* is admirably expressed as a 'carrying asunder'. Perhaps, however, the best illustration is the word *trait*, derived from the Latin *tractum*, meaning something that is 'drawn away'. It exactly describes the isolation of a trait by its withdrawal from the idea-cluster of which it formed part. The appositeness of many analogical extensions shows an acuteness of intelligence which may lead us to believe that they suggested themselves subconsciously. It is probable, indeed, that subconscious thought has contributed very materially to the formation of words—that affinities actuated the brain independently of consciousness. We have seen in chapter v that the formation of complex and abstract ideas evades scrutiny: it cannot be observed, and can therefore only be traced by inference. On *a priori* grounds it is, therefore, likely that subconscious action should contribute to the making of the emblems by which ideas are expressed; and it will have been observed in the course of our discussions that reasoned conclusions that seemed difficult of acceptance have been confirmed by an appeal to the original significations of words—meanings which they bore when first invented, but have since faded into obscurity.

Meanings may be extended associatively so as to express the antecedent (or cause) or the consequent of the idea that was originally signified. Thus *box* is so-called from its material, *damask* from its place of origin, a *cold* because of its cause, a *safe* because of its consequence. *Putrid*[8] and *filth* originated from expressions of disgust, and are applied to their causes. On the other hand, *chaste* and *pure* refer to their purifying consequences: *esteem* means both an actual feeling of respect, and the idea of this feeling which is its consequence.

Restrictions of the original meaning of words have probably been of more importance than is commonly supposed. For it seems that, primitively, an action was not conceived apart from its object, so that a single 'portmanteau' word was used for the verb and its accusative. Survivals of this practice still exist in the dialects of the American Indians. The verb was isolated or differentiated from its object by the fact that it retains its identity when the object is changed.

[8] These associations are copiously illustrated in Wedgewood's *Origin of Language*. It seems probable that *chaste* (*castus*, καθαρὸς) and *pure* express consequences of clearing the throat and of expectoration. The former is from a root *khwad* which recalls the process vulgarly known as 'hawking'.

By the amalgamation of utterances multitudes of new words are formed, as is illustrated by *horsemanship*, in which the final syllable signifies an individualized trait. There are numerous compound adjectives, such as *hateful*, which express a consequence, but are used in a causal sense. The meaning of a word may be modified by the addition of a syllable: *in* or *un* signifies a negative: *dis* (apparently implying *distance*) a contrary: *re* a 'back-lash': *per* (right through) success, *mis* failure—probably from 'midst' in the sense of *incomplete*, as in the slang phrase 'not half' for *quite*. Relationships of circumstance may be signified by case endings that are attached to nouns and adjectives, or by syllables that are amalgamated with the verb, as in *understand* and *forgive* (in which *for* represents *from* and signifies the relationship of inconsistency) and the hosts of verbs that commence with a Latin preposition. And syllables may be employed to denote the time of a verb, and the person for which it is used.

The imitation of sounds is at best imperfect; when an object is symbolized through an associated sound, several sounds often offer themselves for adoption: a movement of the limbs, or a feeling of touch, includes many traits or associations, any one of which may be imitated; the vagaries of playful imagination are infinite. Accordingly primitive vocabularies may assume very different forms, and we can understand why man expresses himself in a Babel of languages. Speaking generally, each word must have been invented by a particular individual,[9] just as are the creations of art, mechanical contrivances, and fashions in dress. It becomes established either because of its intrinsic appositeness or the respect in which its inventor is held, so spreading from the family to the tribe. Within the tribe it is subject to constant modification. For,

[9] We forget our benefactors, and think as little of these originators as of those who revolutionized domestic life by the invention (for instance) of chairs, tables, and window-glass. The word 'gas' is, however, known to have been invented by the chemist Van Helmont.

quite apart from the effect of fancy in suggesting changes, there is a persistent tendency to shorten or soften words in order to reduce the effort of uttering them, leading to such abbreviations as 'don't' and 'bike', and to the dropping of aspirates. Consequently, if tribes originally speaking the same language become isolated from one another they develop languages which appear to be distinct. One of the hill tribes of Assam (the Nagas) has split up into groups of villages, each of which has been at perpetual war with its neighbours, since it has become a mark of social distinction to have taken the heads of men, women, and even children, by incursions or in ambushes: indeed, to be eligible for marriage, a man must have had one head to his credit. It follows that there has been no peaceful communication between one group of villages and another, and, secluded in this fashion, each has elaborated a special tongue of its own. No less than seven distinct languages are spoken by at most 200,000 persons—all radically identical, but so different in form as to be quite unintelligible outside their boundaries.

Generally, however, as tribal isolation has been broken down by the domination of conquest, or the intercommunication of trade, the words and constructions of one language have filtered into another. In this transfusion of language words are not uncommonly so greatly modified [10] that their genealogy becomes obscured. Into the word 'alarm', for instance, the Italian preposition and article have been welded. And it is to be noted that the speech, like the arts, of a cultured people tends to spread amongst its uncultured neighbours unless a change of tongue is opposed by strong national sentiment. For, being associated with higher culture, it is superior and stimulates admiring imitation.

Except when they are proper names or pronouns,

[10] Interesting illustrations of these corruptions are given in Weekley's *Romance of Words*.

words express general ideas of kind and trait—of character or relationship. But, when linked into sentences, they become particularized by their accompaniments. A sentence ascribes traits to an individual. In modern European languages ascription is effected by unifying the individual with a trait-class, by associating it with a trait, or by unifying it with a trait as a *feeling*. But it may be left to be inferred from a succession of words that are grammatically quite unconnected, as in many American-Indian dialects. The traits which are ascribed to the individual are generally *coincident* in their relations to it. But, presenting themselves successively, they are expressed in *sequence*: a sentence expresses a coincidence as a sequence, and since traits may be apprehended in different orders, the arrangement of words in a sentence may vary. We say 'he is ill in bed with a violent cold': but we might also say 'with cold in bed is he—violent'. To us there appears to be a 'natural' sequence—the subject, or pronoun, preceding the verb, the object following it, and qualifications generally preceding the words which they define. But in inflected languages, sentences are composed without regard to conventional order. There is little observance of definite arrangement in classical Latin: the change into mediaeval Latin was marked by nothing so clearly as by the substitution of a modern phrasing for the puzzling inconsequence of the classical style. Moreover, sequences, as we have seen, are reversible, and the order of words may be consciously modified in the interests of emphasis and rhythm.

We introduce traits of relationship into a sentence by the use of prepositions that express the relation of the word that follows them to the subject of the sentence. But both subjects and traits may be expressed by adding syllables or letters to words or by incorporating [11] them in words. In this case words are 'inflected'. The

[11] In Turkish, for example, a negative meaning is given to a word by inserting the letter *m* into it.

syllables or letters appear to have been originally words, signifying the person or time of a verb, the gender of the subject, or the relations which we indicate by prepositions. In some languages these inflections are so complicated and so regular[12] that we can hardly doubt that they were invented or systematized by primitive 'academies' of language.

Words and sentences are symbols with which the imagination can play, and consequently the elaboration of language became a fine art: the selection of words and the construction of sentences offer in themselves subjects for decorative activity. Pleasure is given by the observance of rhythm, and the introduction of alliteration and of rhyming identities of sound. These are the added charms of poetry. Language, like the other arts, comes to us with pleasure in each hand—that of the ideas which it expresses, and that of beauty of technique. We are gratified by that which is expressed, and by that which expresses it. The liking for beauty leads us to avoid words that forcibly recall disagreeable, startling, or indecent ideas, and to substitute for them expressions that are less picturesque[13] in their appeal. So the brain drifted into the vagaries of euphuism

And words are things of power in themselves. By recalling ideas they are active stimuli, with an energy that impresses us. 'Behold, how great a matter a little fire kindleth.' The ideas which they summon may materialize themselves associatively (auto-suggestively), not merely in action of the limbs, but in physical conditions which appear to be beyond the control of the brain, as is shown by the phenomena of hypnotism.

[12] Arabic is provided with a scheme of internal word changes by which the bearing of any root can be uniformly modified, so that a simple notion of action can be systematically converted into one of the agent, effect, and act in the abstract. This elaboration must have been the artificial fruit of purposeful invention.

[13] By a reverse process, forcible 'vulgar' words are constantly being coined to replace words that have lost their picturesque significance. 'Pluck' gives place to 'guts', both having the same meaning.

Being muscular movements they produce the effects of muscular activity: if we curse another we feel that we have injured him. Accordingly words have been treated with great respect. They have been venerated as talismans against ill-luck, and in the East are commonly worn on the necks of children. They have been accepted as of good or ill omen, and those which appeal strongly to the imagination have been personified—even deified. The idea of Liberty has, as we have seen, a real potency. But how many of us worship simply the *word*!

Writing is symbolic, but its symbolism may be direct or indirect. In the former case it signifies an idea, in the latter case, the sound of the word which expresses the idea. That is to say, it may be ideographic or phonetic. To the first class belong all the most ancient systems of writing—the hieroglyphics of Egypt, the cuneiform of Assyria, and the ideograms of Chinese. They are, in fact, pictures, which have evolved precisely as the pictorial art has evolved. Being independent of sound, they have the advantage of overriding language, and may be comprehended by men who speak different tongues if they are acquainted with the system on which ideas are pictured. But they are cumbrous, they cannot without great difficulty be modified so as to express the grammatical relations of words to other words, and their use involves two wholly different sets of symbol-ideas for writing and speaking. It was simpler, and therefore pleasanter, that writing should symbolize spoken symbols, than that it should involve the employment of a wholly distinct set of signs.

There was, then, a conscious motive for the development of phonetic writing, as soon as the conception of expressing a sound by a mark had arisen. This could come into being by a simple reversal. A mark is a thing: things are expressed by sounds, and sounds may, therefore, be expressed by marks. The mind, as we have seen, may reverse any connexion in sequence. Hence the written symbol of one idea could be employed to represent another idea which was expressed in speech by the same sound. For the development of phonetic writing it was necessary to analyse words into syllables, and syllables into letters. This was effected by the same process of isolation as that by which a sentence is analysed into parts of speech. A consonant is isolated from a vowel because it can be sounded with any of them, or can be pronounced soundlessly. It can therefore be figured as a distinct individuality. The signs that were adopted for these elements of sound were in some cases derived from the ideograms of words which were taken to typify them: consonants were more easily represented in this fashion than vowels, and accordingly, in some languages, such as Arabic, the alphabet is practically limited to consonants, vowels being expressed by supplementary marks, set above or below them. But the ancient Ogham characters of Ireland attempted to signify letters by dots and dashes, such as are now used in the Morse system of telegraphing.

The adoption of phonetic writing tended to stereotype spoken language by fixing word-sounds. An unwritten language, or one which uses pictographic characters, is constantly changing, as new utterances are invented, or established utterances are simplified in the interests of easy speaking. But when sounds are symbolized in writing there is a resistance to evolutionary change. It is not, however, altogether effectual, and in many countries the vocabulary of everyday speech differs radically [14] from that employed in writing.

[14] Educated Arabs will deny that some of the commonest (and most ancient) words in use *can* be written, since they are not included in the classical vocabulary.

NOTE E

In the brain, as we have seen, ideas of utterances lie close against those of finger-movements. They are normally situated in the left hemisphere, but shift to the right hemispheres of those who are left-handed. And movements of the finger and thumb are so closely analogous to those of the lips and the tongue that they may be used to classify the consonants as finger-signs. This is attempted in the table opposite.

In forming 'breathed' consonants, movements are unchanged, but they fall short of *touching*. H is formed by a contraction of the chest muscles, and is analogous to a 'grip' or 'hitch'. In uttering vowels there may be touched or breathed movements of the glottis, which in Arabic are signified by separate letters.

Utterances may be complicated very greatly by the association of two or more consonants in a sequence. With some limitations any consonant may be followed by H, or by a *press*, *jerk*, or *rub*. X is the CH of *ich* followed by S: Q is K followed by W. S may lead almost any consonant or sequence of consonants.

That it should be possible to represent the leading consonants so exactly as movements of the fingers and thumb seems to show that the uses of the hand and the mouth, as instruments of expression, evolved together, the latter finally monopolizing the field as having infinitely greater capabilities. In some cases finger-gestures closely imitate those of the arm and hand: for this reason, indeed, can they appropriately be described by the names which we have given them. Can any other analogies be traced between them and actions, such as would have inspired sign-making by an appeal to subconscious intelligence? A press—or a rub—implies resistance, and hence a negative may be expressed by N, or, as in Arabic, by L or M. A jerk is clearly energetic, and the finger-movements for *wake, win, joy,* and *yes* are therefore emblematic. There is also significance in the 'rubbed' finger-signs for *lie, lick,* and *loll.* We should need fancy to discover any inherent significance in the movements classed as 'tappings'. Moreover, the distinction between 'sharp' and 'flat' (as between *tap* and *dab*) became obscured as the use of the tongue developed, and 'breathed' consonants evolved. Accordingly sharp, flat, and breathed consonants of each kind in this class could supplant one another as dialects evolved into languages.

Movements of the fingers and thumb.	*Consonantal utterances of the mouth.*						
	Lips as fore-finger and thumb.		*Fore-tongue as second finger, palate as thumb.*		*Hind-tongue as second and third fingers, palate as thumb.*		
	Touched.	*Breathed.*	*Touched.*	*Breathed.*	*Touched.*	*Breathed.*	
Tapped together :							
—by their ends	P	F	T	TH (*thin*)	K	C	sharp
—by their first joints	B	V	D	TH (*this*)	G	CH (*loch*)	flat
Pressed together	M	FF	N	S(lisped TH)	N (*bien*)	CH (*ich*)	—
Jerked apart :							
—from their ends	—	W	CH (*child*)	CH (*chère*)	Z	ZH (*jeu*)	sharp
—from their first joints			J	Y			flat
Rubbed together	—	—	L	R	LL (*ville*)	R (*rue*)	—

It is noticeable that *presses* and *rubs* are distinguished in their alphabetical names by being introduced by a vowel e.g. eM instead of Be.

CHAPTER XXI

SUGGESTION

THE young of the higher vertebrates are not left to pursue independently their conscious evolution. Their parents communicate to them by gestures the knowledge and habits which they have themselves acquired or learnt from their progenitors. Evolution is, accordingly, assisted by an 'accumulated' environment. The invention of speech immensely increased the range of this environment, presenting to each individual the conclusions, beliefs, and prejudices, as well as the inventions and habits of his predecessors. These are easily memorized, and the evolution of the individual is accelerated by the assimilation of the knowledge and thoughts of his society. Progress would, then, have advanced with ever increasing rapidity, had it not been arrested by prejudices, and thrown back by natural catastrophes and by war.

We employ the term 'suggestion' for the process by which the acquisitions of a society are transferred to the individuals that are born into it. The thoughts of one man, signified by his words, set others thinking with him, and call up ideas that would otherwise not have presented themselves. The current of his reflections induces a similar—or, it may be, a reversed, or negative—current, in the minds of those who comprehend him. An audience listening to a speaker is in nervous sympathy with him: their eyes follow his gestures, their ears receive his words, and their brains form the connexions and disconnexions of ideas that constitute his thoughts. And, since these ideas stimulate psychic energy, they also *feel* with him. Accordingly the process of suggestion involves the active condition that is involved in *comprehension*.[1] One comprehends another by forming the idea-connexions that are symbolized by his words. If his words are not understood in this fashion, they do not translate themselves into thoughts. If they are understood, they may not merely recall thoughts that have been previously formed, but lead to the formation of new thoughts by the establishment of new idea-connexions.

By suggestion members of the same community exercise upon one another a power which is really tremendous. Man, as a solitary, could advance but little by resisting his physical propensities, by asserting himself against his environment, or by the conscious development of new tastes. But the suggestions of another affect him as stimuli, and serve as stepping-stones to progress. And he again may similarly lead the steps of his fellows. Suggestion links men together into a social organism, the cells of which act upon, and are affected by, those surrounding them.

Words are so much more explicit than gestures that speech has almost entirely superseded gesture-language. But for deaf-mutes gestures are the only means of communication available, and have been elaborated into the complexity of a language. When the deaf are also blind, gestures must actually touch them to be perceived; yet, so delicate are the associations which the sense of touch can form with the brain that (as shown by the amazing talents of Helen Keller) tactile impressions alone can suffice to impart suggestions which will supply ample material for the formation of thoughts. Persons of very delicate tactile sensibility, who can by self-abstraction exclude any disturbing thoughts or sensations, are able, in contact with another, to comprehend involuntary muscular[2] indications of direction and

[1] The signification of which is expressed in the idiom—'grasping his meaning'.

[2] When Helen Keller was holding the hand of her friend and instructress, she appeared to recover her sense of hearing, and would turn her head and smile when others spoke or whistled. Away from

purpose which earn for them a mysterious credit for 'thought-reading'. And, however eloquent be a speaker's words, they are effectively reinforced by his gestures and facial expressions. Indeed by them alone an assembly to whom an orator's words are unintelligible[3] may be inspired with enthusiasm for the cause which he is advocating. They follow his emotions. But they cannot, of course, follow his *thoughts*, unless they understand the language which he employs.

We may listen or read critically or uncritically. In the former case the brain effects the connexions indicated by the words to which we are attending, but it permits contrary thoughts to present themselves, and these, if strong enough, may reverse suggested connexions into negatives. Such a statement as that 'the heart of the people beats true to our policy' may be reversed by recollections of strikes, and 'beats' becomes 'does not beat'. If our comprehension is uncritical, the brain forms without question the idea-connexions that are suggested. It is uncritical if the words call into action a prejudice which enforces them, or if the speaker or writer can engage our confidence or faith by commanding our respect. But it may happen, later on, that the brain, in the course of its activity, will draw forward thoughts which contradict our biassed or trustful conclusions.

Suggestion is expressive or expectant according as it simply manifests the pleasure or displeasure of ideas, or is dictated by a practical object. The former includes speaking or writing which is playful, artistic, amusing, or which simply expresses praise or blame with no direct purpose of influencing the opinion of another. It affords a like pleasure or displeasure to him who comprehends it in a sympathetic spirit. Its features have already been discussed in our consideration of expression, and

her friend, she remained motionless, not giving the least sign that she realized what was going on. (*Story of my Life*, p. 354.)

[3] Of this St. Bernard's preaching of the Crusades is a striking illustration.

we need do no more here than refer to it. Our concern in this chapter is with expectant or practical suggestion —that which is made 'with an eye to business'. Its objects are manifold, and it may be distinguished as interrogative, informative (or indicative), argumentative, imperative, persuasive, and deceptive. A practical spirit may enter into suggestion that is primarily expressive, such as poetry. In this case art becomes didactic.

First of *interrogative* suggestion. One who puts a question to another suggests to him a doubt that is in his own mind—it may be as to the other's thoughts, intentions, or conduct, it may be as to the associations or character of something that is obscure. He may at the same time suggest a possibility which would solve the doubt, as in such a question as—'Are you going for a walk?' or 'Is it an aeroplane?' If the doubt conflicts with a conclusion which the other has formed, its suggestion produces a reversal of cerebral excitement that is expressed by a negative reply. If it is not reversed in this fashion, it stimulates an effort of thought which may resolve the doubt through a familiarity of association (rhythm) or trait. One decides to take, or not to take, a walk because of associations; one concludes that an object is, or is not, an aeroplane from its traits.

By *informative* (or indicative) suggestion civilization is preserved: the stores of knowledge that have been accumulated by past generations are handed down to their successors. Knowledge may be defined as the mass of familiarities of trait or association that have been consciously established up to date between different ideas and link them together—all the percepts and thoughts, which predicate of things that they *are* other things, or that they *have* other things as their antecedents, accompaniments, or consequences. They may have been established by reasoning, as well as by perception. Reasoned conclusions may, as we have seen, be formed subconsciously, in which case we become aware of them before their reasons present themselves. But in explaining them to another their reasons must be stated, and comprehended,[4] if they are to be assimilated by him, and not merely committed to memory, as is the case when they are accepted as articles of faith.

Faith is no satisfactory substitute for the process of reasoning. But it is our only anchor-hold for much of the associative experience that is used in reasoning—of experience, that is to say, of sequences, or of causes and consequences. It is impossible to verify for ourselves all the sequences that have been observed. When they are the past occurrences which are recorded in history, reason can only employ itself in examining the credibility of our authorities. And there is a limit to the possibility of verifying sequences which are capable of being verified: all the experiments of chemistry cannot be repeated in the class-room. But when the reason of an argument is based, not upon association, but upon a sameness of trait, the unification must actually take place in the pupil's mind if the argument is to be comprehended, and not merely remembered. Unifications or classifications that are accepted upon trust afford no exercise to the reasoning faculties. And if the events recorded in history are not woven into sequence by being referred to general laws, we learn nothing of the future from the past and its chapters merely afford such amusement as can be derived from a novel, or become an instrument for implanting certain dogmas of faith.

Suggestion becomes *argumentative* or polemic, when it employs reasons in order to reverse the established conclusions of another—that is to say, to convince him of error. If, however, he holds these conclusions by

[4] Comprehension is facilitated by an acquaintance with more than one language, since thoughts can then be isolated from particular memorial attachments, and it is realized that they exist independently of words.

prejudice or faith, he has no ears for the lessons of experience, and accordingly reasoned argument may be less useful in disseminating truth than in exercising the intelligence of him who employs it.

Imperative suggestion is command, the exercise of authority over others. This is delightful to every man, since it associates him with an idea of power. It is accepted in obedience when it is received in faith—the nervous condition which is produced by the idea of protective power. Reason incessantly conflicts with faith, endeavouring to undermine its authority. But faith remains the ruling guide of life. It may be discredited as a means of civil government. But it rules in the nursery and the schoolroom—the convent and the army—and in every disciplined association of mankind. And those who decry it as an instrument of statecraft, willingly submit themselves to it in their social and religious life. A nation without faith is adrift on the waves of class antagonism. For reason cannot guide it, as a whole, by the recollection of consequences, since the interests of all classes are not alike, so that consequences that will be advantageous to some will be harmful to others. There is, for instance, a standing opposition between the producer and the consumer, between the capitalist and the workman, which can only be quieted by faith in the government which stands between them. If faith could be extinguished by experience, there would be little hope that national coherence would endure. But all that reason can destroy is the association of faith with a particular individual or authority. In itself faith is a nervous reaction that is inevitably stimulated—even in the most sceptically disposed—by authority which is really powerful and protective.

Persuasion is perhaps the most interesting form of suggestion, since it assumes such diverse phases. It may be figured as a four-armed goddess—acting upon us by appealing to common sense with reasons that may be either true or deceptive, by working upon our feelings or by kindling our imagination. Its instrument is the pleasure or displeasure which it promises. Faith is indifferent to advantage or disadvantage, but persuasion moves us through the prospect of good or evil.[5] The relationship of persuasion to command is, therefore, the same as that of choice to emotional impulse. Persuasive inducements may be the reasoned consequences of the course which we are invited to adopt or warned to avoid; in this case they are based upon experience. But they may be fanciful consequences which may lure us into error. Advertisements appeal to us by the offer of advantages which may be actual or imaginary, by the skilful use of temptations, which is the art of the commercial traveller and is illustrated in the devices by which salesmen play upon the judgement of their customers.

We are, however, most easily persuaded when the appeal is made to our feelings—to hopes or fears, likes or dislikes, prejudices or partialities. So we are exhorted to buy goods of 'British make'. A candidate for parliamentary honours can rely more safely upon personal or party popularity than upon any arguments that he puts forward in his electioneering address. Persuasion by prejudice is the instrument that is used by a government to produce the national uniformity of antagonism that must be maintained in war time. No argument is too fatuous to be accepted when it is supported by a strong dislike. In the days when we hated Napoleon it was hardly a joke that he 'filled the butchers' shops with large blue flies'.

Persuasion appeals to the imagination when it takes a *rhetorical* form—when its inducements are presented, not as reasoned arguments, but as lively images which appeal to the perceptive rather than the reflective faculties, and are impressive because perception is more

[5] As a child may be induced to go to bed quietly by the promise of a sugar-plum or of a threat of the 'black dog'.

vivid than thought. Their vivacity energizes speech with the eloquence which is so persuasive. And rhetoric makes large use of ideals. These 'carry us off our feet' in a whirl of imagination excitement. They may be fantastic. But they may be of immense practical utility. For it is a reassuring fact that men are almost as easily touched by appeals to kindness [6] and humanity as to their self-assertive or malevolent feelings.

If one desires that another should think with him, he must realize the effect of his words by identifying himself with his hearer or reader. This is sympathy, without which few writers or speakers can be convincing. Accordingly, one who thinks for purposes of suggestion has an imagined personality before him, realizes the effect of his ideas upon it, and hastens to counteract any that may be displeasing. He disarms jealousy by self-depreciation, elaborately excuses any views that may give offence, anticipates doubts and objections and is ready with replies to them. And he remembers, above all things, that emotion is stronger than intelligence, and that if he can hold the attention of his audience, and enlist their prejudices, he can persuade them 'against themselves'. Their attention will be held if the subject is one which excites strong like or dislike or a lively curiosity. Failing this, its presentation must be brightened by flashes of humour, or by force or eloquence of language. This, being admired in itself, may suffice to hold an audience spell-bound.

We prefer persuasion to command because it flatters our self-esteem. It is not inconsistent with the independence of liberty that we should follow another's suggestions for doing what we please. Accordingly in

[6] As, for instance, by the ideal that it is the bounden duty of each generation to promote the healthfulness, intelligence, and morality of the generation that is to succeed it—a duty to which even vested interests must pay obeisance. It ordains the ruthless elimination of city slums and the careful purification of popular entertainments. It is admired as a sentimental but unpractical aspiration. Yet an earnest appeal to it would generally command a large majority vote. And in America few would dare argue against it.

a democracy persuasion is enthroned as the Genius of the State. Politicians and journalists educate themselves to be her priests: indeed they may be termed, not unfairly, 'professional persuaders'. But their appeals must generally be illogical and inconsequent, since, except under a socialistic régime, the interests of different classes of the community do not coincide: consequences which would benefit one section would be harmful to another, and hence reasoned arguments cannot convince a nation as a whole. Political persuasion must therefore rely upon prejudices, or upon the enthusiasm which it can kindle for ideals. But prejudices change, and ideals are evanescent. In the long run reason triumphs; and a community which is not welded together by loyalty for a strong authority tends to divide itself up into syndicates or castes, each united by a similarity of interests, and each striving for its own advantages.

But, it will be urged, whatever be the merits or demerits of democracy, it has procured for us the boon of free speech, for this is the extension of liberty from the persuaded to the persuaders. It is, indeed, a potent evolutionary force, since it offers an unfettered range for the influence of suggestion. It is attractive, since it represents self-power in expression, and men have endured persecution and even death on its behalf. Consequently liberty of speech has become a popular ideal, which may be respected by an absolute government. But, so far as appears from history, its security depends upon party spirit—upon the competition between different classes of politicians to obtain the direction of State affairs. For, if the class in power can stifle criticism, their antagonists would have no instrument but force wherewith to supplant them. When, therefore, party spirit ceases, as during war, speech may be brought under the severest restrictions—indeed, freedom of expression may be denied even in the writing of private letters.

Finally, we arrive at *deceptive* suggestion, the object of which is to mislead another. The primitive idea-connexions of deceit are so simple as to be within the capacity of quite small children. They are reversals that are occasioned by fear—by an apprehension of blame, punishment, or failure—which are the consequences of thoughts that are in another's mind, and also in our own. The effect of apprehension upon these thoughts is to reverse them, just as fear will reverse an idea of enterprise. 'I did do it' becomes 'I did not do it' in the same fashion as 'I will jump it' becomes 'I will not jump it'. We may then conclude that a child's first essays in deceit are subconscious, and are not deliberately directed to mislead another. But by an intelligent appreciation of advantageous consequences deceit becomes an instrument that can be deliberately employed—an expedient which may be used in affirmative as well as negative suggestion, in order to induce another to fall in with our views—to accept, for instance, an article of which we wish to dispose, or a conclusion which suits our purpose.

This view of the origin of deceit may appear to be far-fetched. It is, however, confirmed by experience of childish behaviour. Untruthfulness commonly goes with timidity, and the earliest falsehoods of infancy appear to spring to the lips uncontrollably. But upon this simple foundation expectative thought builds up innumerable contrivances by appreciating samenesses of trait that present to it, as its materials, the infinite varieties of speech and conduct. For deceit may be as effective a weapon as force. Current morality reprobates it strongly, and it is certainly less dignified than open antagonism, or the self-discipline of confession. But, judging from ancient literature, it has not always been in disfavour. The Hebrew scriptures attribute it, dispassionately, to exemplary characters; and Athéne could find a good reason for her patronage of Ulysses in her admiration for his skill in lying speciously. It is, however, fatal to trade. This rests upon credit, and it appears that the development of commerce has contributed to an increasing appreciation of truth, by demonstrating that it possesses great economic advantages.

CHAPTER XXII

CHARACTER AND EDUCATION

Our inquiries have led us to the simple but profound conclusion that a living creature is a nervous system to which organs are attached for the nourishment and organic stimulation of the nerves, and muscles for the discharge of their energy in movement. Nerves are instruments for the transformation of stimuli into living activity, and nervous excitement must be preceded by a stimulus—instinctive, associative, or unificative—unless there is to be a consequence without a cause. The excitement varies in degree according as the nerves are more or less susceptible to the stimulus. It follows that the individual peculiarities called 'traits of character' result from something that is peculiar in the sensibilities of individuals to particular stimuli.

These peculiarities are inborn, although their consequences may be modified by experience and education. We come into the world differing from other individuals in our sensitiveness to stimuli that touch the physical, psychic, or reflective branches of the nervous system, and affect our behaviour irrespective of volition. The effects of these stimuli are accompanied by pleasure or displeasure, which sway our behaviour by influencing the course of choice. Their intensity, again, primarily depends upon our inborn sensibilities, but it may be increased by practice. From them are evolved more complicated 'tastes' which are artificially associated with certain stimuli and become objects of choice. They may be intensified by practice. But they are ultimately traceable to inborn sensibilities, and hence it is rightly held that, while our voluntary behaviour is modified by experience and education, our character comes to us by birth.

The sensibilities on the physical plane, of which we are aware, are attended by consciousness, but exist independently of it. They are obviously more compelling in some individuals than in others. In terming men greedy, sensual, or slothful we admit that they are abnormally affected by the organic stimuli which manifest themselves in the appetites. It is manifest that some men are touched more acutely than others by the neuro-sensory stimuli of fear. Emotional susceptibilities differ. The effect of purely sensory stimuli probably varies from one man to another much more considerably than we suppose. We cannot compare the intensity of our own sensations with that of another's, except by its perceptible consequences. These assure us that some persons are colour-blind—others quite insensible to the charms of music. Some are abnormally sensitive to peculiar stimuli: there are those who have an exaggerated horror of snakes; others are uncomfortable if a cat is in the room. Certain conditions of nervous weakness are accompanied by a curious dislike of open spaces: the patient suffers from the peculiar susceptibility which makes a wild animal reluctant to break cover. If conduct is any guide to feeling, we may be sure that some men are much less sensitive than others to the pressures which cause physical pain.

The effects of physical susceptibilities are reflected on the psychic plane as states of exhilaration and depression—of joy and sorrow. But they may also be *resisted* unconsciously or subconsciously, and the intensity of both reflections and resistances will depend upon the sensibility of the individual's psychic energy and its strength. If, as appears, psychic energy is the result of nervous stimulation, its strength will again be the result of nervous susceptibility. Psychic strength manifests itself in phlegmatic determination:

psychic sensibility in an excitable or enthusiastic temperament. One, or the other, is a very prominent trait of character. Being less acutely sensitive to unfavourable conditions, the phlegmatic do not suffer a marked nervous contraction when exposed to them: they resist them easily, and are curious without enthusiasm, courageous with *sang-froid*. But, since there is less contraction, there is also less of the succeeding expansion which in consciousness is pride; and it is in every one's experience that the phlegmatic take their struggles against the unfavourable as a matter of course, are far less elated by success than those who have won it through a strong nervous reaction, and occupy themselves far less than the excitable in the pursuit of the stimuli of pride. In fact they take life philosophically.

The excitable, again, fall into two classes—those whose temperaments are elastic, and rapidly recover from a nervous shock, and those with whom nervous irritability tends to become chronic. This is the difference between optimism and pessimism. The optimist may be exceedingly sensitive to rebuffs and disappointments. But he throws off their effects, and, in the process, experiences acutely the pleasure which is caused by nervous revulsion. Consequently the joy of pride is to him an irresistible attraction. The pessimist, on the contrary, is haunted by the fear of shame, since in his case nervous contraction is not overbalanced by succeeding expansion. Both one and the other pursue the same objects. But they are actuated by very different spirits—the one by hope, the other by fear. Both may practise self-control, act courageously, or seek the esteem of others. But the optimist is led by aspirations, the pessimist by apprehensions, and their characters are consequently very different. The optimist is vain of his appearance, conceited of his talents, arrogant of honour: the pessimist is disposed to be touchy, shy, and reticent, and his fear of the disapproval of others produces in him a nervous respect for convention.

Psychic resistance stimulates an effort which is consciously manifested in *willing*. The will borrows its materiality from the physical plane, but arises as a spiritual impulse. Accordingly will-power, whether assertive, tentative, or selective, depends upon psychic strength, or, as it is termed, 'strength of character'. Independence of thought, the verification of conclusions, promptness of decision mark psychic forcefulness: by the possibilities of choice the weak are thrown into a tremor which can only be stilled by another's advice.

Peculiarities of sensibility on the reflective plane—those which affect the working of the brain—are appreciated by us through their effects in producing various degrees of intelligence and of memorial capacity. Intelligence is susceptibility to familiarity of trait: memory susceptibility to familiarity of rhythm. Both are far more acute in man than in the lower animals, and between man and man there are similar differences, although of vastly less degree. We have seen that muscular movements may be guided by familiarity of trait, as well as by memorial association. Imitation involves the expression of an idea by a movement which possesses a resemblance to it. Dexterity is, then, not wholly memorial, and in admiring it we are attracted by something more than deft automatic repetition. There is, it appears, an intelligence of body as well as of mind.

When the working of the brain is affected by psychic excitement, it becomes imaginative. The degree in which it is so influenced is a marked trait of character. A vivid imagination implies psychic excitability. Sensitiveness is the foundation of the artistic temperament, but its effectiveness depends upon intelligence and memory. Unintelligent imagination is merely hallucination. In its infancy, imaginative activity is Play: arriving at maturity, it flowers as Art, and bears fruit as Religion.

The imagination, as we have seen, affects voluntary

behaviour by tinging it with idealism, romance, or self-consciousness. The effect of the two former upon character is sufficiently obvious. But we may advert again to the influence of self-consciousness. This is the consciousness of other selves that are within us—images that are formed by the personification of our physical and psychic traits. To the self-conscious man temptation involves a conflict between physical and psychic personalities, appreciated by the personality of his brain. He can, therefore, watch his psychic self struggling against his physical self, and is moved by a desire to win the respect of the latter by overcoming it, and thus to gain the pleasure of pride. He is, then, haunted by a contrast from which there is no escape, and can be trusted to observe the rules of his morality when there is no one but himself to appreciate his conduct. He is imaginative. But his imagination has a practical effect upon his behaviour.

The generation of these inborn susceptibilities appears to be affected in some degree by environal influences. We constantly refer to 'national' or 'racial character'. The traits which we attribute to other peoples, as their *instinctive* possessions, are for the most part merely acquired peculiarities—the results of peculiar experiences and education, acting upon the individual after birth. But in some cases they appear to have a deeper foundation. Southern nations are generally more emotional and artistic than Northerners, and we may infer that an instinctive sensuousness and excitability is developed by warmth and brilliancy of surroundings. On the other hand there is more strength and elasticity of temper in the North than in the South, and each of us knows well, from his own experience, that cold is favourable to the development of nervous tone.

The responses of all these inborn sensibilities may be reinforced by indulgence, since nervous action is facilitated by repetitions that establish associative connexions. And they are powerfully controlled by the

effects of the pleasures and displeasures that are their conscious accompaniments, and by the artificial tastes that are evolved from them. These affect us in a threefold manner. They may lead to manifestative or 'sentimental' behaviour, which is an end in itself. By influencing choice they may be the motives of the *practical* conduct which is a means to an end. And, through the imaginative excitement which they arouse, they may infuse practical behaviour with a 'sentimental' complexion, rendering it idealistic, romantic, or self-conscious.

The pleasures that have been distinguished as 'tastes' are, as we have seen, of a highly complicated nature. In those of the aesthetic kind, physical sensibilities combine with psychic ideas of excellence: economic pleasures contain physical and reflective elements. But, being ultimately based upon inborn susceptibilities, they respectively depend upon a natural sensuousness or practicalness of disposition, and vary in strength from one individual to another. One who has no eye for colour cannot appreciate painting; one who lacks ear for music can derive no pleasure from it. Economic pleasures are dull to those in whom psychic enthusiasm overrides practical purpose.

Pleasure and displeasure are particularized by being associated with particular stimuli. The association is natural when it is instinctive: we naturally associate pleasure with those whom we love. But most associations are artificial, conventional, or symbolic, and are brought about by experience and education. There is nothing instinctive in the association of beauty with a particular picture,[1] of dignity with a particular decoration, of amusement with a particular game. But associations are in some cases influenced by inborn sensibilities. The instruments for procuring pride and

[1] The style of picture, money, or amusement which stirs the aesthetic, economic, or psychic sensibility of a Chinaman generally leaves a European unmoved. But the kind of taste to which they appeal is the same.

avoiding shame, for instance, vary in attractiveness with differences in physical or emotional susceptibilities. They may be aggressive, self-repressive, accumulative, industrious, pretentious, or solicitous, according to the character of the individual. So again with amusements. To appreciate the ludicrous requires intelligence. But all men are amused by the fortuitous and the mysterious.

Speaking generally, however, pleasures are particularized by experience and education, independently of character. Experience includes the influence of the environment, and this may accentuate some tastes and weaken others. The attractiveness of economic pleasures—of the amassing of wealth, or of the profits of industry—has, as we have seen, grown with the use of money. And it is noticeable that the craving for the esteem of others is a peculiar feature of town life, apparently evolved under social pressure. The baron in his castle, the cultivator on his farm, regard public opinion with indifference, and are but little amenable to political persuasion.

Intellectual pleasures are given by the expansiveness of curiosity in itself, by the pride which comes of successful exploration, or by the ideal of truth for its own sake. Curiosity, as a form of subconscious resistance, depends upon psychic strength and sensibility: there is no one who is not affected by it in some degree, but, failing resistant power, it is only aroused by the commonplace. Explorative zeal requires intelligence for its satisfaction, and this results from a sensibility of the brain. But it is nerved by the resistant energy of assertive independence. Failing this, one seeks guidance from faith. The self-confident, who are sufficiently excitable to feel the shock of doubt, fight it, not with the armoury of faith, but with their own weapons—that is to say, by reason—possessing, as it is aptly phrased, the 'courage of their own opinions'.

It is a fact of momentous importance that pleasures of all kinds, like the appetites, are strengthened by

indulgence—that is to say, by practice. They become familiar, and familiarity enhances their attractiveness. Accordingly, by education and practice, man can enhance the influence of stimuli, and may infect himself with likings for pursuits that would naturally leave him unmoved. So he comes to *take* pleasure in the study of science and philosophy.

Let us now turn to Education. This is a peculiar phase of experience—that which is gathered under the influence of faith. It commences in the nursery, and is continued until we are launched into the world to learn the truth, not by faith, but by reasoning influence. Faith is strongest during childhood, since the need of protection is, then, constantly evident. And during childhood memorial susceptibilities are at their acutest. It is, accordingly, the education of the first four or five years of life which contributes most masterfully to our culture. This we owe to our mothers.

The acquisitions which we derive from education are dexterities, habits, tastes, knowledge, and beliefs. They come to us in some measure by unificatings that are made by the brain—that is to say, intelligently. But, for the most part, they are memorial associations. We think of dexterities as capacities for skilled imaginative expression—such as the playing of games or of musical instruments. But, in reality, they include everything in action or utterance that distinguishes man from the apes. Eating, speaking—and even upright walking, must be learnt, and the difficulties to be overcome in so ordinary a process as dressing oneself can be realized if one attempts to reverse the folding of a necktie. Little children learn the complicated movements of manners and speech with extraordinary facility. As years advance, there is a loss of plasticity: a language is rarely pronounced and accentuated correctly unless it is spoken in childhood. For the same reason the practising of craftmanship must be begun in early youth: to acquire an art, schooling must give way to

apprenticeship. Eloquence is a dexterity, the usefulness of which grows as persuasion displaces command in the controlling and influencing of others. The Greeks recognized its value, and taught it systematically. But its artistic cultivation finds no place in our school and college *curricula*—perhaps because it is derogatory to liberty to realize that it can be blinded by artifice.

Habits are memorial associations, as is implied in their name.[2] They may be of mind or of body. Habits of mind are associations of ideas, such as those which unify us with a particular nation or class, give us faith in a particular religion, government, or institution, and form our tastes in dress and in art. They are also of immense importance in endowing symbols with significance, and so creating a vast number of artificial stimuli. It is by a habit of mind that the national anthem becomes inspiring. Habits of body may be disciplinary —that is to say, they may enforce self-control by facilitating its exercise. Discipline, in its narrowest sense, signifies the control of spasms of self-confidence which lead to the disobeying of authority. But, taken broadly, it includes all the self-repression that gives decency and order to life—from the restraint of bodily promptings that are inopportune, to the curbing of manifestations of anger or jealousy. Good manners involve more than this—the use of conduct and speech that will actually give pleasure to others. At the outset such acquired manners will be artificial and forced. But they tend to become natural, since, through associative, or auto-suggestive, stimulation, actions which express a feeling recall the feeling:[3] hence kindly conduct actually leads to kindly thoughts, and respectful manners to feelings of reverence. 'Manners makyth man.' So effort becomes the easier the more often it is

[2] They are *associated* with us as are our clothes.

[3] We have already noticed that by deliberately smiling or frowning we can recall, however faintly, pleasurable and displeasurable feelings.

practised: courage, as is well known, is fostered by a life of hardship.

Habit is mechanical and spiritless. But by the acquisition of new tastes we summon new motives to our assistance, and can modify our conduct in spirit as well as by rule. It is their complexity[4] that differentiates civilization from barbarism: their transmission by education is what we mean by 'culture'. Economic pleasures need no cultivation: they are sufficiently fostered by the potentialities of money. Nor is any special training required to increase our appreciation of the pleasures of pride or of amusement: education merely attaches them to certain symbols. Aesthetic and intellectual tastes, on the other hand, need forcing. The former include an appreciation of literature, painting, sculpture, architecture, music, and dancing. It seems unfortunate that school-training should generally have concentrated itself so exclusively upon literature. For the other phases of art add immensely to the happiness of life, and tastes that are not acquired in youth are often not acquired at all.

Literature has been so engrossing a study because it has been associated with dead classical languages. One can understand that at the time of the Renaissance Latin and Greek were prized as the seed of a civilization that had been trampled out. But since those days a new culture has sprung up, and languages have evolved into a new form in which inflexions are generally discarded. That the classics should be learnt as an aid to expression is as though horsemanship were included in a course of motor-driving instruction. And, if it be urged that the beauties of classic art can only be appreciated through a study of Greek and Latin, it may be replied that these beauties are not appreciated by the vast majority of those who learn these languages,

[4] A striking illustration is the cult of exercise. To a savage it is incomprehensible that the muscles should be exerted for the pleasure of exertion. Nor can he appreciate the pleasure of gardening.

that there is much that is demoralizing in them, and that modern [5] literary art offers attractions which are of equal excellence and of higher moral value. And, after all, the fact remains that, although language may be studied as an art in itself, it is fundamentally an instrument, and that from this point of view, the most fruitful of language-studies is that of one's own.

In the interests of human happiness there are some imaginative ideals which require more culture than they receive. Ideals which personify attributes of power, such as those of Dignity and Liberty, spring to life of themselves. But Magnanimity and the Christian ideals of human Fellowship and Unity are more refined in quality, and must generally be inculcated to be appreciated. They are difficult to teach. For the young are uncompromisingly logical,[6] and are quick to perceive differences between the principles and practice of those who preach to them.

By 'knowledge' is meant the mass of classifications, associations, and sequences that have been established by the reason and experience of past generations. It includes history, science, and mathematics—that is to say, an acquaintance with the phenomena of the human, animate, and inanimate world, of the laws that can be deduced from them, and of the functions of rhythm and of rhythmic relationships. These fulfil their purpose when they are used as materials and instruments for reasoning thought. The facts and generalizations which they include may provide attractive themes for the imagination. But they are profitable, from the evolutionary point of view, only when they are used as materials for reasoning. We turn to this purpose an

[5] John Bright was one of the most eloquent of men, but had no classical training.

[6] A lady of my acquaintance, who had taken trouble to urge the Christian ideal upon her children, was met with the question, 'But, mother, you have two sets of furs and poor Mademoiselle (the governess) has none. Why do you not give her one of yours?' For very shame she had to do so.

infinitesimally small proportion of the knowledge which we possess. For, since education relies upon faith, it is apt to check the development of reasoning capacity. It is noticeable that little children will reason with much more freedom than their elder brothers and sisters.

It is because education does not teach the *use* of knowledge in reasoning that its results are intellectually so disappointing. The arguments which are successfully employed in journalism and in political propaganda shake one's confidence in the efficiency of the schooling which has been so energetically promoted during the past half-century. And it is remarkable that so many of the men who have led the world by the discovery of new truths have had no educational advantages whatever.[7] They excelled, not because they possessed great knowledge, but because they could use the knowledge which they possessed by drawing deductions from it.

We shall, indeed, hardly exaggerate in concluding that most of the knowledge that is taught to the young is presented to them as a tissue of beliefs, to be accepted upon authority, not to be assimilated. This would involve *thinking*, and, although most schoolmasters would unhesitatingly affirm that boys should be 'taught to think', they show by their practice that they appreciate its dangers. For religious creeds will hardly withstand the test of common sense: all ideals cannot be justified logically; and respect can reasonably be accorded to power only when the power is real. Reason is iconoclastic—that is to say, it is antagonistic to faith and imagination. If faith were eliminated we should ridicule as meaningless much that we hold to be venerable. And if imagination were suppressed, there would be a grievous loss of happiness. Evolution might weave its complexities more rapidly. But this would be a poor consolation.

Without, however, adopting reason as the sole

[7] Kepler spent his early youth as a labourer. Bunyan and Burns, Watt and Stephenson had no regular schooling.

criterion of the good, education would probably do well to neglect it less completely. Its use can be encouraged by giving reasons for conclusions—by setting out the steps by which they have been formed, instead of inculcating them as articles of faith. There are difficulties. Many of the conquests of science are really imaginative and cannot be proved by experience; and the explanation of truths that are demonstrable would add very greatly to the labour of teaching. But if the discoveries of astronomy, for instance, are taught as dogmas, they do no more to train the judgement than the study of folk-lore. And if reason was introduced into educational courses, we should bridge the gulf that divides school and college life from that of business; and young men would not feel, in going into the world, that they met a strange atmosphere, in which values are changed, and their learning counts for little by the side of common sense.

EPILOGUE

THE views to which these inquiries have led us are disillusioning—lamentably disillusioning, many will say—and, were it clear that mankind lived happily in the shadow of self-ignorance, one might shrink from piercing the canopy, and leading in the sunlight. But the briefest examination of the past and the present shows that man wastes the happiness of his talents. Nature offers him many elementary pleasures, from which he can extract a multitude of zests, as a prism draws colours out of daylight. He craves for feelings of superiority and power. But these come to him with advancing years from the gradations of domestic and social life: they may be won by friendly emulation, by overcoming difficulties and by mastering himself. He can amuse himself in a hundred fashions, and is supplied with aesthetic playthings by imaginative Art. Religion—and even philosophy—may lift him to an ecstatic plane from which earth's pleasures look sordid, and its pains contemptible. He is blessed with inspiring ideals of magnanimity and sympathy. Yet, between the pompous lines of history, one can read a miserable tale of pain and suffering—the perennial fruits of war, of intolerant persecution, and of the oppression of the helpless in actual or economic servitude. We are, indeed, coming to realize that history is a picture-film of torturings, enlivened by a military band. If we regard the present, we find the horrors of war increased many-fold, and less relieved by the ennobling spirit of chivalry. We boast of our economic civilization; but a walk through the slums of a commercial town will impress us with such degradation as is hardly to be found amongst the poorest classes of the East. And, in our private lives, how much happiness do we not throw away through an insensate sense of our dignity, through the jealousy which corrodes the brightness of friendly intercourse! To realize that these feelings have no particular nobility of origin will not prevent their arising; but it will enable us to check their manifestations in conduct, to limit their forcefulness, and to shorten the periods of their control. It is something to realize that anger, pride, and jealousy are merely nervous reactions, and can be resisted—that economic ambitions are evolved from instinctive promptings, and run contrary to the magnanimous. All said then, it cannot be claimed that illusion is successful; and there is a case for bringing the mainsprings of our behaviour under the light.

INDEX